INCREASING ACCESS TO CLEAN COOKING IN THE PHILIPPINES
CHALLENGES AND PROSPECTS

FEBRUARY 2021

ASIAN DEVELOPMENT BANK

ADB

© 2021 Asian Development Bank
6 ADB Avenue, Mandaluyong City, 1550 Metro Manila, Philippines
Tel +63 2 8632 4444; Fax +63 2 8636 2444
www.adb.org

Some rights reserved. Published in 2021.

ISBN 978-92-9262-695-2 (print); 978-92-9262-696-9 (electronic); 978-92-9262-697-6 (ebook)
Publication Stock No. TCS210018-2
DOI: http://dx.doi.org/10.22617/TCS210018-2

The views expressed in this publication are those of the authors and do not necessarily reflect the views and policies of the Asian Development Bank (ADB) or its Board of Governors or the governments they represent.

ADB does not guarantee the accuracy of the data included in this publication and accepts no responsibility for any consequence of their use. The mention of specific companies or products of manufacturers does not imply that they are endorsed or recommended by ADB in preference to others of a similar nature that are not mentioned.

By making any designation of or reference to a particular territory or geographic area, or by using the term "country" in this document, ADB does not intend to make any judgments as to the legal or other status of any territory or area.

Please contact pubsmarketing@adb.org if you have questions or comments with respect to content, or if you wish to obtain copyright permission for your intended use that does not fall within these terms, or for permission to use the ADB logo.

Corrigenda to ADB publications may be found at http://www.adb.org/publications/corrigenda.

Notes:
In this publication, "$" refers to United States dollars.
ADB recognizes "Vietnam" as Viet Nam.

Cover design by Kris Guico.

Contents

Tables, Figures, and Boxes

Tables

Figures

Boxes

Foreword

The role of the Asian Development Bank (ADB) in advancing the goals of the Sustainable Energy for All (SEforALL) initiative in Asia and the Pacific includes facilitating opportunities for growth of investments that will speed up the sustainable transformation of the world's energy landscape. Through knowledge and information exchange, leveraging of existing energy structures, and mobilizing and consolidating efforts of development partners, it is hoped that a more conducive policy environment will be developed to further accelerate the development of a more ecologically sustainable energy future.

The report *Increasing Access to Clean Cooking in the Philippines: Challenges and Prospects* was prepared in keeping with ADB's continuing efforts to maximize its support to the global aspiration of energy for all and to strengthen its investments and increase its project portfolio in the area of energy access under its Energy for All (E4ALL) Initiative. Moreover, as the leading partner and host of the Asia–Pacific SEforALL hub, the publication of this report signifies in concrete terms ADB's support to the goals of SEforALL and that of the United Nations' Sustainable Development Goal 7 (SDG 7) of universal access to modern, affordable, reliable, and sustainable modern energy for all.

Specifically, the report documents the results of intensive research, survey, and investigation of Philippine cooking practices with the aim of providing an on-the-ground perspective of the issues and potential solutions relating to access to clean cooking. Through this report, ADB also hopes to add to the knowledge about the impact of continued use of traditional cookstoves that use charcoal and fuelwood to households in terms of exposure to household air pollutants, and the associated challenges to be hurdled in order to institute the desired changes and contribute to the achievement of SDG 7 and other associated SDGs by 2030.

The conduct of this research on access to clean cooking is an initiative of the Sector Advisory Service Cluster-Energy Sector Group of ADB to promote SEforALL in Asia and the Pacific. The research is part of the Cluster TA 0017 (REG): Promoting Sustainable Energy for All in Asia and the Pacific under TA 8946: Energy Access for Urban Poor (Subproject B), which aims to provide accessible, cleaner, and more efficient energy in Asia and the Pacific through assisting selected developing member countries in identifying, designing, and developing projects that will address the unmet energy needs of the urban poor, and preparing projects and programs for replication and scaling up to other developing member countries.

The study findings and the key takeaways would serve to encourage and guide stakeholders—energy policymakers, national and local government units, civil society organizations, the private sector, clean cooking investors and developers, donors, and financing institutions—to design and implement appropriate interventions that can lead to increased access to cleaner cooking technologies and fuels.

Robert Guild
Chief Section Officer
Sustainable Development and Climate Change Department
Asian Development Bank

Acknowledgments

ncreasing Access to Clean Cooking in the Philippines: Challenges and Prospects is a product of extensive field research, survey, and investigation into the household cooking practices in the Philippines carried out by the Sustainable Development and Climate Change Department (SDCC) of the Asian Development Bank (ADB) through technical assistance 8946: Energy Access for Urban Poor (Subproject B) under cluster regional technical assistance: Promoting Sustainable Energy for All in Asia and the Pacific. The study was conducted by a team in the Sector Advisory Service Cluster–Energy Sector Group (SDSC-ENE) led by Kee-Yung Nam, principal energy economist. Yongping Zhai, chief of the Energy Sector Group and Robert Guild, chief sector officer of the SDCC provided overall guidance. A group of international and national experts provided invaluable contributions as authors of background papers.

This report was written by a team of experts from the SDSC-ENE, under the guidance of Kee-Yung Nam and the supervision of Yongping Zhai. The team comprised Yun Ji Suh, energy specialist; Felicisima Arriola; Mylene Cayetano; Elmar Elbling; Denise Encarnacion; Lyndree Malang; Marcial Semira; Ana Maria Tolentino; Maria Fritzie Vergel; and Grace Yeneza. Charity Torregosa, senior energy officer; Maria Dona Aliboso, operations analyst; and Angelica Apilado, operations assistant provided technical advisory and administrative support. Ma. Theresa Mercado copyedited the report, Kris Guico designed the cover, and Mike Cortes prepared the layout.

The report also benefited from insights and comments of ADB colleagues from the energy divisions of ADB's regional departments. The views and opinions expressed here are those of the authors and do not necessarily reflect those of ADB, its governors, or the governments they represent. The study and publication of this report is courtesy of the Government of Austria.

Abbreviations

ADB	Asian Development Bank
AP-SEforALL	Asia-Pacific Regional Hub of the Sustainable Energy for All Initiative
AQI	Air Quality Index
CNG	compressed natural gas
CO	carbon monoxide
COPD	Chronic Obstructive Pulmonary Disease
DMC	developing member country
DOE	Department of Energy
ERT	emission rate target
ESMAP	Energy Sector Management Assistance Program
FAO	Food and Agriculture Organization of the United Nations
GACC	Global Alliance for Clean Cookstoves
GHG	greenhouse gas
HAP	household air pollution
HAQ	household air quality
HUC	highly urbanized city
ICS	Improved Cookstoves
IEA	International Energy Agency
IHME	Institute for Health Metrics and Evaluation
IRENA	International Renewable Energy Agency
LGU	local government unit
LPG	liquefied petroleum gas
NAAQGV	National Ambient Air Quality Guideline Values
NCD	noncommunicable disease
NO_2	nitrogen dioxide
$PM_{2.5}$	atmospheric Particulate Matter that have a diameter of less than 2.5 micrometers
SDG	Sustainable Development Goals
SEforALL	Sustainable Energy for All Initiative
SO_2	sulfur dioxide
UN	United Nations
UNDP	United Nations Development Programme
UNESCAP	United Nations Economic and Social Commission for Asia and the Pacific
UNSD	United Nations Statistics Division
US EPA	United States Environmental Protection Agency
WBT	water boiling test
WHO	World Health Organization

Weights and Measures

°C	degree Celsius
$\mu g/m^3$	micrograms per Cubic Meter of Air
g	gram
kg	kilogram
kWh	kilowatt-hour
L	liter
mg	milligram
m^3	cubic meter
mg/m^3	milligram per Cubic Meter
MWh	megawatt-hour
tCO_2e	tons of carbon dioxide equivalent
Tj/ton	terajoule per ton

Executive Summary

Replacing traditional methods of cooking using open fires and solid fuels with clean cooking solutions is an integral element of Sustainable Development Goal 7 (SDG 7), which aims to achieve universal access to modern, affordable, reliable, and sustainable modern energy for all. Reliance on inefficient cooking practices amplifies household air pollution (HAP). This practice brings about serious health and environmental consequences that impact about 4 million people who die prematurely each year from illnesses attributable to HAP. The Energy Sector Management Assistance Program (ESMAP) and the World Bank estimated in a 2020 report that inaction in meeting the 2030 targets costs the global economy $2.4 trillion annually, with the health impact alone accounting for 58.3% of this cost. Thus, HAP does not only affect health but also have far-reaching implications to development, affecting the overall goal of the 2030 Agenda for Sustainable Development.

Despite efforts to increase global access to clean cooking, the *Tracking SDG 7: Energy Progress Report 2020* points to 2.8 billion people still without access to clean cooking as of 2018. Some 1.8 billion of these people live in Asia and the Pacific, per the United Nations Economic and Social Commission for Asia and the Pacific (UNESCAP) Policy Brief. The slow progress in the deployment of clean cooking solutions highlights the need for identifying more specific interventions that would appropriately address the gaps at the country level.

The same 2020 SDG 7 tracking report has identified the Philippines as the country with the slowest progress, at only 8% from 2000 to 2018, in access to clean cooking among countries in Southeast Asia. While it is well-known that the use of traditional cooking fuels is a leading cause of household air pollution, the extent of this pollution and its impact on the health and environment of communities in the Philippines is not well established. This lack of information could be the reason for the apparent inattention given to the ensuing health and environmental issues from traditional cooking practices. Considering this situation, this study was therefore undertaken in the Philippines, to gather information on the current household cooking practices and to determine the impact of a potential shift to improved cooking technology and fuel use on indoor air pollution. The study will augment available information and knowledge about the extent of air pollution due to traditional cooking practices in the Philippines and foster understanding of the prevailing barriers and issues relative to increasing access to clean cooking. The results of the study could serve as constructive inputs to the development of specific policies to address the country's slow progress in increasing access to clean cooking.

The Philippine experience, which focuses on Iloilo City—representing a highly urbanized coastal city, and San Jose City—representing a low-density peri-urban and landlocked city, highlights the implications to households if access to clean cooking cannot be promoted effectively. While a variety of improved and modern fuels and cooking technologies are already available in local markets, traditional cement stoves utilizing either charcoal or fuelwood are still widely used either exclusively or in combination with gas stoves utilizing butane or liquefied petroleum gas (LPG) stove, and electric stoves. The 2020 SDG 7 tracking report estimates access to clean cooking in the Philippines at 46% of its 2018 population with rural areas lagging at 27%. This leaves some 54% or about 57.6 million people relying on traditional cement cookstoves and utilizing charcoal or fuelwood as cooking fuel.

The field surveys in the two study sites revealed that 46.8% (Iloilo City) and 16% (San Jose City) of households exclusively use traditional cookstoves. Only approximately a quarter of the population for both cities (25.4% in Iloilo City, 27.5% in San Jose City) exclusively use modern stoves that are either fueled by butane, LPG, or powered by electricity. Moreover, due to the prevalence of fuel stacking practice, 25.4% of households in Iloilo City and 55.5% of households in San Jose City still use traditional alternately with gas or electric cookstoves.

Field emission test results show that the prevailing traditional fuel–technology combinations used by most households are found to be unsafe health-wise, at point of use. When compared with the emission rate targets (ERT) based on the World Health Organization (WHO) *Guidelines for indoor air quality: household fuel combustion,* the emission level measured using water boiling tests (WBT) for traditional cookstoves with either charcoal or fuelwood greatly exceeded the WHO ERT for $PM_{2.5}$ and CO. Likewise, the emission concentrations within households during field emission tests when compared with the United States Environmental Protection Agency (US EPA) Air Quality Index (AQI) showed that traditional cookstoves using charcoal or fuelwood emit pollutants at levels that are very harmful to the health of every household member regardless of age group or gender. These findings were further confirmed by laboratory tests which revealed very high emission concentration levels of particulate matter ($PM_{2.5}$) and carbon monoxide emitted by traditional cookstove-fuel combinations.

From controlled WBTs conducted in the laboratory, thermal efficiencies of the cookstove and fuel combinations for traditional cookstoves using charcoal or fuelwood, LPG stove, and electric stoves were also obtained. The laboratory tests found that, among the stoves tested, electric and gas stoves are the most efficient with 33.4% and 26.5% thermal efficiency, while traditional stoves utilizing charcoal or fuelwood were very inefficient at 5.2% and 10.4% thermal efficiency, respectively.

The thermal efficiency measurements also provided estimates of time and fuel needed to heat a specific quantity of water, which were then used to estimate the costs that households incur with their current cooking practice and food preferences. Estimates show that charcoal is by far the most expensive among the four cooking fuels, costing households ₱18,414.01/year ($346.16) in Iloilo City and ₱15,800.10/year ($297.02) in San Jose City. This is followed by modern cookstoves using LPG, which can cost households in Iloilo City ₱6,893.73/year ($129.59) and in San Jose City ₱4,813.62/year ($90.49). For households utilizing electric stoves, a household in Iloilo City can incur additional electricity cost of ₱3,838.44 ($72.16) or ₱3,314.35 ($62.31) in San Jose City per year. The least expensive cooking fuel is fuelwood, with households estimated to spend only from ₱958.83 ($18.02) in Iloilo City to ₱151.78 ($2.85) in San Jose City per year.

Switching from traditional cookstoves, using either of the two traditional fuels, to gas or electric stoves will lead to significant reductions in $PM_{2.5}$ emission concentrations especially for households exclusively using traditional cookstoves. A switch from traditional stoves using charcoal can decrease $PM_{2.5}$ emission from as low as 60.04% (for a shift to butane stove) to as much as 99.32% (for a shift to electric stove). A switch from traditional cookstoves utilizing fuelwood as cooking fuel, to modern cookstoves, can decrease $PM_{2.5}$ emission by at least 84.38% (for a shift to butane) to as much as 99.74% (for a shift to electricity). Of the households exclusively using traditional cookstoves, 54% in San Jose City and 73% in Iloilo City indicated their willingness to shift to clean cooking. The main barriers that hold back these households from switching to clean cooking solutions include: (i) the up-front costs of stoves, and the recurring costs of fuel, either LPG or additional electricity charge; and (ii) the perception that cooking using traditional cookstoves is more convenient, faster, and are safer than the unfamiliar modern cookstoves.

The findings and key takeaways point to policy, information, technology and financing gaps as well as prospects to foster access to clean cooking in the country. With additional facts and learnings gained from the study, energy policymakers, local government units, clean cooking investors, and other stakeholders may be encouraged and moved to formulate, design and implement country- or situation - specific policies and programs to fast-track market expansion and the switching to more efficient, cleaner cooking technologies. A better understanding of the issues and challenges in the access to clean cooking space would also enable the Asian Development Bank to assess where it could contribute knowledge and resources in support of clean cooking access efforts not only in the Philippines but also in other developing member countries across The Asia and Pacific region.

1. Introduction

According to the World Health Organization (WHO), household air pollution (HAP) is the "single most important environmental health risk factor worldwide." HAP is often caused by the still pervasive use of polluting fuels such as charcoal, cokes, fuelwood, or agricultural wastes for cooking, lighting and heating. These inefficient cooking practices produce high levels of air pollutants such as particulate matter ($PM_{2.5}$), Carbon monoxide (CO), sulfur dioxide (SO_2), and nitrogen dioxide (NO_2)—exposure to which have been associated with various health concerns.[1]

It is estimated that 4.3 million of the 7 million premature deaths due to air pollution each year are from illnesses attributable to HAP, which include noncommunicable diseases (NCDs) such as stroke, ischemic heart disease, lung cancer, and chronic obstructive pulmonary disease (COPD). Furthermore, HAP disproportionately affects the world's most vulnerable—putting women, children, the elderly, the displaced and the extremely impoverished population at a higher risk of disease from exposure (footnote 1). In 2020, ESMAP and the World Bank estimated that the costs to the global economy of meeting the 2030 targets of universal access to clean cooking amounts to $2.4 trillion annually, with the health impact alone accounting for $1.4 trillion or 58.3% of this cost. Also included in this cost estimate are the costs to climate ($0.2 trillion) and gender ($0.8 trillion).[2]

The ultimate goal of the United Nations (UN) is to end all forms of poverty and hunger, protect the planet from degradation, ensure prosperous and fulfilling lives for human beings, and foster peaceful, just, and inclusive societies.[3] One of the 17 Sustainable Development Goals (SDGs), i.e., SDG 7, is to "ensure access to affordable, reliable, sustainable and modern energy for all." This goal has five major targets, namely: (i) universal access to modern technology; (ii) increase global percentage of renewable energy; (iii) double the improvement in energy efficiency; (iv) promote access, technology and investments in clean energy; and (v) expand and upgrade energy services for developing countries (footnote 3).

Universal access to clean cooking, along with universal access to electricity, is an integral element of the SDG7 target of universal access to modern technology to be achieved by 2030. ESMAP, in 2020, identified these clean cooking solutions from the health perspective; among these are liquefied petroleum gas (LPG), electricity, improved cookstoves (ICS) such as best-in-class gasifiers, biogas digesters, and solar cookers (footnote 2).

---

[1] WHO. 2016. Burning Opportunity: Clean Household Energy for Health, Sustainable Development, and Wellbeing of Women and Children. Geneva. p. 130.

[2] Energy Sector Management Assistance Program (ESMAP). 2020. *The State of Access to Modern Energy Cooking Services.* Washington, DC: World Bank. License: Creative Commons Attribution CC BY 3.0 IGO

[3] UN General Assembly. 2015. Transforming our world: the 2030 Agenda for Sustainable Development. 21 October. A/RES/70/1.

The importance of universal access to clean cooking cannot be overemphasized. HAP, brought about by the use of inefficient cooking, not only impacts health but also affects other factors of development such as poverty, gender inequality, environmental degradation, air pollution, and climate change. Increasing access to clean fuels and technologies can therefore greatly contribute to the achievement of 10 out of the 17 SDG Goals.

In the latest *Tracking SDG 7: Energy Progress Report*, 2020, global access to clean cooking is reported to have increased from 56% in 2010 to 63% in 2018 leaving an estimated 2.8 billion people worldwide without access to cooking systems.[4] Of these, 1.8 billion or 64% live in Asia and the Pacific.[5] While there has been some progress, the pace is not sufficient to achieve the universal access target by 2030. It is also apparent that this reported increase in access to clean cooking has not cascaded evenly across all countries worldwide. As reported, the top 20 countries with the largest populations lacking access to clean cooking fuel and technologies accounted for 82% of the global population without access between 2014 and 2018. The Philippines is listed among these 20 countries.

Access to clean cooking in the Philippines is at 46% of its population in 2018 with rural areas lagging at 27% (Table 1). This leaves some 54% or around 57.6 million people relying on traditional cement cookstoves and utilizing charcoal or fuelwood as cooking fuel. Compared with other countries in Southeast Asia such as Indonesia and Viet Nam, the pace of increase in clean cooking access in the Philippines has been much slower. The lack of information on the extent of HAP and its impact on health and environment of communities in the Philippines could be a reason for the apparent inattention given to the ensuing health and environmental issues from traditional cooking practices, resulting to this relatively lackluster performance.

Table 1: Comparison of Clean Cooking Access of Southeast Asian Countries, 2018

Southeast Asian Countries	Percent of Population with Access in 2000[a]	Percent of Population with Access in 2018[a]	Percent Increase in Access to Clean Cooking (2000–2018)[b]	2018 Population Without Access (million)[c]
Brunei Darussalam	>95	>95	no change	<1
Malaysia	>95	>95	no change	<2
Singapore	>95	>95	no change	<1
Thailand	63	79	16	14.6
Viet Nam	13	64	51	34.4
Indonesia	6	80	74	53.5
Philippines	38	46	8	57.6
Myanmar[d]	<5	28	<24	38.7
Cambodia[d]	<5	22	<18	12.7
Lao People's Democratic Republic[d]	<5	6	<3	6.6

[a] International Energy Agency, International Renewable Energy Agency, United Nations Statistics Division, World Bank, World Health Organization. 2020. *Tracking SDG 7: The Energy Progress Report.* Washington DC.
[b] Values are computed from % of population with access in 2000 and 2008.
[c] Computed from percent of population without access in 2018 against World Bank 2018 population data from World Bank. 2020. World Development Indicators (Population, total). Last updated 1 July 2020. Accessed 6 August 2020.
[d] Percent of population with access in 2000 were approximate values so percent increase was also approximated.

[4]   International Energy Agency (IEA), International Renewable Energy Agency (IRENA), United Nations Statistics Division (UNSD), World Bank, WHO. 2020. *Tracking SDG 7: The Energy Progress Report.* Washington, DC.

[5]   IEA. 2019. Clean cooking access database. Accessed 28 April 2020.

A policy brief released by UN in 2018 identified three major barriers or challenges to efforts of transitioning toward universal access to clean cooking. These include: (i) supply issues or the lack of clean, affordable, and available supply of clean fuel and energy sources; (ii) demand issues which include cost of clean fuel and/or device, consumer preference and practices, and overall awareness; and, (iii) enabling environment or the existing monetary and fiscal policies that restrict or inhibit sector growth and sustainability whether due to lack of funding, poor implementation, or poor cross-sectoral coordination.[6]

With energy policymakers, clean cooking technology investors, and other stakeholders as the intended audience; this report aims to provide, through the Philippines' context, a perspective on

(I)   current fuel–technology combinations that households employ, their efficiency and the effect on indoor air quality;

(II)   the costs involved in the utilization of current fuel–technology combinations and the impact of switching to clean cooking on these costs; and,

(III)   the barriers and possible solutions to switching from traditional, inefficient stoves and fuel to clean cooking.

The study will also augment available information and foster understanding of the extent of HAP and the prevailing barriers and issues that hinder deployment of clean cooking technologies and serve as invaluable inputs to finding viable solution to increasing access to clean cooking. From the output of the study, the Asian Development Bank (ADB) can assess where it could contribute knowledge and resources in support of clean cooking access efforts not only in the Philippines but also in its other developing member countries (DMCs) across the Asia and Pacific region.

The first three chapters provide the Philippine context on cooking practices and preferences that define prevailing fuel–technology combinations and how these affect indoor air quality and household health. It investigates the degree of pollution that household members are exposed to. This information is vital in order to generate more appreciation of the issue of indoor air pollution by local governments and their constituents. The next chapters focus on the barriers, including the costs, and the prospects in switching to clean cooking. The report proceeds as follows:

Chapter 2 situates the reader to the Philippines' context. A broader context on fuel use and cookstove preferences are initially presented. This is followed by the results of household surveys that shows current and site-specific cooking practices and local preferences that influence prevailing fuel–technology combinations.

Chapter 3 discusses the results of field emission tests, done in conjunction with the household surveys. The water boiling test (WBT) method was adopted for field emission testing to provide actual field measurements of the amount of air pollutants ($PM_{2.5}$, CO, NO_2, SO_2) emitted by the various fuel–technology combinations employed by households in the study sites. These emission rate measurements for $PM_{2.5}$ and CO were compared with the WHO *Guidelines for indoor air quality: household fuel combustion,* which set standards for clean burning in the homes. Emission concentration of $PM_{2.5}$, CO, NO_2, SO_2 were also compared with the United States Environmental Protection Agency (US EPA) Air Quality Index (AQI), which provided comparisons between ranges of emission and its possible consequences to health.

---

6   UN. 2018. Accelerating SDG 7 Achievement: Policy Briefs in Support of the First SDG 7 Review at the UN High-Level Political Forum. *Policy Brief #2: Achieving Universal Access to Clean and Modern Cooking Fuels, Technologies and Services.*

Chapter 4 presents the results of laboratory tests conducted and its implications to the efficiency of various fuel–technology combinations. Laboratory tests conducted also employed the WBT method, with the controlled environment allowing for comparison among results not only of the emissions but also in terms of the thermal efficiency of the different fuel–technology combinations. The thermal efficiency tests provided controlled estimates of time and fuel needed to heat a specific quantity of water. This also allowed for the estimation of costs involved using household survey results of cooking practice duration.

Chapter 5 discusses the perceived barriers to shifting to clean cooking technologies and assesses the prospect of such a shift. This chapter further estimates the costs and emission reduction effect of switching from exclusive use of traditional and inefficient cooking practices to cleaner solutions. Due to the practice of fuel stacking, the impact on those employing a combination of cooking modalities was not determined, although, it may be deduced that some emission reduction can be achieved, if households switch to the use of improved cookstoves.

Chapter 6 concludes with the major findings, key takeaways, and recommendations formulated from the Philippines cooking study experience.

2. Typical Filipino Cooking Practices and Access to Clean Cooking

The Philippines, composed of 7,641 islands, is an archipelagic country with a population estimated at 108.1 million as of 2019, 51.2% of whom reside in urban areas.[7] Culturally diverse, the various colonial influences, faith-based customs and limitations, and indigenous traditions are evident in its food. Traditional Filipino dishes are generally simple, local cuisines may differ based on regional location and dominant agricultural produce in each area. Viands for meals are usually a combination of fish or meat and vegetables cooked with broth. Dried and fresh fish are pan-fried in oil or grilled over firewood or charcoal. Most if not all meals revolve around the staple steamed rice. As with any typical Filipino household, women are still more often relegated with the task of preparing these meals.

The local market offers a variety of cookstoves for households from traditional cooking technologies such as the traditional cement stoves that can be used together with either firewood or charcoal, or modern cookstoves utilizing clean fuels such as LPG or butane, or electricity. Improved cookstoves (ICS) that allow for continued use of biomass as fuel but offer a more efficient cooking experience are also available, though not as extensively as the traditional and modern stoves (Box 1).

In the latest available census of household energy consumption in 2011, the most commonly used cooking fuel is fuelwood (54%), followed by LPG (40.5%), charcoal (35.3%), and biomass residue (20.1%), which include agricultural and forest products residue. Electricity, kerosene, and biogas complete this total.[8] According to the latest Philippine Forestry Statistics (2018), Philippine wood production is estimated at 999,000 cubic meters of which 2.7% are used as fuelwood, 23.9% are processed into charcoal, and the remaining 73.4% are logs processed into other wood products. Fuelwood and charcoal that remain in the country to supply local demand is estimated at 265.9 cubic meters, while only around 11 cubic meters are exported.[9] Production of charcoal to supply local demand is usually done via traditional methods; that is, by using earth kilns that are considered environmentally degrading, not to mention inefficient.[10] The continued production of charcoal contributes to environmental degradation in the Philippines where it remains a significant source of indigenous energy.[11]

[7] World Bank. 2020. World Development Indicators (Population, total). Last updated 1 July 2020. Accessed 6 August 2020 and Government of the Philippines, Philippine Statistics Authority. 2015. Census of Population and Housing: Highlights on Household Population, Number of Households, and Average Household Size of the Philippines. Manila.

[8] Government of the Philippines, National Statistics office and the Department of Energy. 2011. *Household Energy Consumption Survey 2011*. Manila.

[9] Government of the Philippines, Department of Environment and Natural Resources. 2018. Philippine *Forestry Statistics, 2018*. Forest Management Bureau: Manila.

[10] Ortwien, Andreas and Militar, Jeriel G. 2015: *Use of Biomass as Renewable Energy Source in Panay. Final report*. Manila, Philippines: Deutsche Gesellschaft für Internationale Zusammenarbeit (GIZ) GmbH.

[11] Inzon, M.R.B.Q., M.V.O. Espaldon, et.al. 2016. Environmental Sustainability Analysis of Charcoal Production in Mulanay, Quezon, Philippines. *Journal of Environmental Science and Management. 2016*: 93–100.

Box 1: Traditional, Improved, and Modern Cook Stoves available in the Philippines

Traditional cookstoves pertain to either open fires or cookstoves, usually made of cement, with a wide base opening to accommodate biomass fuel such as fuelwood, charcoal, biomass residues, dung, etc. These stoves, which are usually constructed by artisans or household members are considered to have poor combustion features and therefore energy inefficient.

In the Philippines, the simplest of traditional cookstoves commonly feature open wood or charcoal fires underneath pots supported by steel rails (photo 1) or firewood and charcoal clay or cement stoves such as shown in photo 2. Outdoor cooking and grilling is also done using grillers similar to that of photos 3 and 4. On the other hand, modern cookstoves available in most Philippine appliance stores are those that operate with the use of fuel contained in a canister or tank, such as butane canisters and liquefied petroleum gas (LPG) tanks (photos 5 and 6), or those that use electricity to produce heat (photos 7 and 8).

Traditional cookstoves. Pictures of traditional and modern cookstoves used in the Philippines.
Photos 6 and 7 from field survey, photos 1,2,3,4,5,8 from www.shutterstock.com.

Improved cookstoves (ICS) pertain to cookstoves that still use traditional charcoal or solid biomass fuel such as fuelwood but have been developed and equipped with improved physical features that can facilitate better combustion, therefore improving cooking efficiency, and decreasing fuel use that would otherwise lead to more pollutant emissions. ICSs offer an alternative or an upgrade from the traditional cookstoves. However, these are still not widely used, and the manufacturing of these products is not yet considered an industry. Some of the improved cookstoves developed by local cookstove manufacturers in the Philippines include the following:

Mabaga kalan   **Biolexis**   **Wonder kalan**   **Papa brick stove**

Improved cookstoves in the Philippines. Photos of available improved cookstoves in the Philippines.
Photo by Elaine Arnaiz in Dubois, M., C. Roth and C. Talamanca (ed.), 2017. *StovePlus Academy 4th Edition: Business Development for Improved Cookstoves and Innovative Fuels.*

1. The Mabaga Kalan is a charcoal stove that claims to save up to 60% of charcoal consumption compared to traditional stoves. This improved cookstove is made from cement and galvanized iron. It is said to be smokeless and features an insulator which compresses heat.
2. The Biolexis Multifuel Gasifier Stove is a portable stove which can operate using wood chunks, wood shavings, charcoal, coconut shell, corn cobs, nut shells, rice husk, among others. It was developed with the primary purpose of utilizing free and abundant "waste" resources for fuel to reduce or even completely eliminate expenses on fuel. It is said to be smokeless and its waste product called "char" can be used as organic fertilizer for plants.
3. The Wonder Kalan is an improved biomass cookstove that operates on charcoal. The stove fire power can be regulated through a vent that can be opened or closed as needed. The product claims to be safe to use and economical, able to promote fuel savings.
4. The Papa Brick Stove is a gasifier made of ceramic; bricks in its combustion chamber are constructed in sections to avoid cracking from intense heat. It utilizes Pili nut (*Canarium ovatum*) shells as its fuel. Because of the hardness of the Pili shell this leads toward a slow start, but it remains hot for a longer time and is ideal for slow cooking of Filipino meat and soup dishes.

Source: World Bank. 2011. Household Cookstoves, Environment, Health, and Climate Change: A New Look at an Old Problem. Washington, DC; Biolexis Multifuel Gasifier Stove; StovePlus Academy 4th Edition; Business Development for Improved Cookstoves and Innovative Fuels, 2017; Guinto, J. 2015. Anatomy of the PapaBrick Stove. September (unpublished). Accessed 10 December 2019.

Local supply of charcoal and fuelwood are mainly used by households for cooking and food preparation. The latest available demand data however indicates that for fuelwood, only 15.3% of Philippine households purchase it from the markets while 79.3% are self-collected or gathered and the remaining do a combination of self-collection or purchasing fuelwood. For charcoal, this is reversed with 92.2% of households purchasing it from the markets while the remaining percentage of households do a combination of self-collection and purchasing for their consumption.[12]

As earlier stated, access to clean cooking in the Philippines was only 46% by 2018, or an increase by only 8% from 2010 to 2018, leaving around 57.6 million of the population without access to clean cooking (footnote 4). Tracking this progress is not easy as there is little existing documentation on the cooking sector and no updates on government policies regulating cooking technologies or promoting access to clean cooking. In its publication, *Energizing Finance: Taking the Pulse 2019*,[13] SEforALL estimated that clean cooking access in the Philippines may increase to 53% or to 13 million households by end of 2018. However, this SEforALL report includes the population practicing fuel stacking modern cooking technology with traditional cookstoves, which it estimates to be around 6 million or 46% of these households with access to clean cooking (footnote 13).

2.1.   The Study Sites

To be able to attain the objectives set out for this study, two focus areas in the Philippines were identified: Iloilo City in Western Visayas, which is a coastal, highly urbanized city (HUC), and San Jose City in Central Luzon, which is a landlocked peri-urban component city. Previous census data have indicated that these cities relied heavily on traditional cookstoves that utilized charcoal and/ or fuelwood as cooking fuel.

Iloilo City has a land area of 7,834 hectares, a coastline area of 21.3 kilometers, and total population of 447,992.[14] It is the regional capital and administrative center of Western Visayas with sea- and air- port linkages to Metro Manila and other major growth centers in the country as well as to some international destinations.[15] With a population density of 57.2 persons per hectare, it is also the fifth densely-populated HUC in the country.

San Jose City, on the other hand is a third-class component city of the province of Nueva Ecija, the province which produces most of the country's rice supply. San Jose City is less densely populated compared to Iloilo City, with its population of 139,738 people spread over 18,725 hectares or a population density of just 7.5 persons per hectare.[16] As a peri-urban city its population and income satisfy the requirements for its city classification, but agriculture remains as the main source of livelihood of its populace. San Jose City differs from Iloilo City not only demographically but also geographically. It is an inland city located within the vast plains of Central Luzon and easily accessible from Manila through a few hours of land trip. Figure 1 shows the location of the two study sites.

---

[12]   Government of the Philippines, Philippine Statistics Authority and Department of Energy. 2011. *Household Energy Consumption Survey*. Manila.

[13]   Sustainable Energy for All (SEforALL) and Catalyst Off-Grid Advisors. 2019. *Energizing Finance: Taking the Pulse 2019*. Washington, DC.

[14]   Government of the Philippines, Philippine Statistics Authority. 2015. *Census of Population and Housing*. Manila.

[15]   Government of the Philippines, Local Government of Iloilo City Official Website of Iloilo City.

[16]   Government of the Philippines, Philippine Statistics Authority. 2015 Census of Population and Housing; Household survey. Manila; ADB. 2015. Promoting Sustainable Energy for All in Asia and the Pacific - Energy Access for Urban Poor. TA 8946.

Figure 1: Philippine Map and Inset of the Two Study Sites

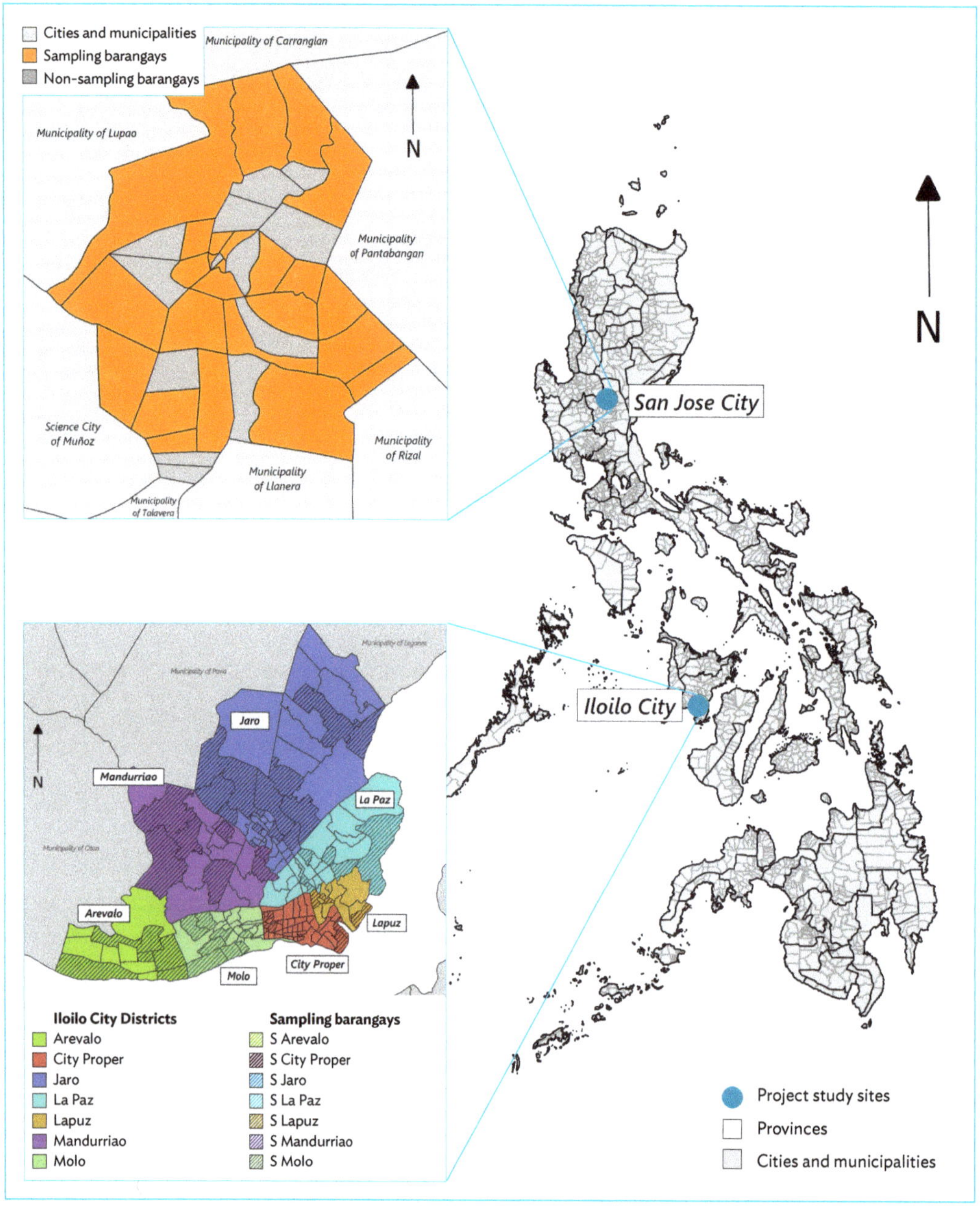

Source: ADB. 2015. Promoting Sustainable Energy for All in Asia and the Pacific – Energy Access for Urban Poor. TA 8946. *Surveyor's Manual, 2018.*

The household survey began on 5 October 2018 and concluded on 7 December 2018. Stratified random sampling was employed in selecting respondents which covered 201 households from 50 of 180 barangays in Iloilo City and 200 households from 25 of the 38 barangays in San Jose City. The survey explored sample households' kitchen structures to evaluate ventilation levels, cooking practices, and food preferences, and choice of cookstove and cooking fuel combinations.

2.2. Cooking Habits and Practices

For the purpose of the survey, the types of food prepared by households were loosely categorized by the manner that these are cooked. Food are either oil-based or fried, water-based such as steamed rice or soups, grilled, or smoked. Rice, as well as some other water-based viand are prepared most frequently as indicated by 65%–85% of respondents followed by fried food. Grilling meats, which is popularly attributed among people in the Visayas region was surprisingly not the most common way of preparing food in households according to the results of the survey though the survey process was not able to delve into the respondents' reasons for this. The least-preferred cooking method among respondents is smoking, which was practiced by only some respondents in San Jose City. Table 2 presents a summary of these results.

Table 2: Cooking Habits and Practices of Survey Households in Iloilo City

| Types of meal by manner of cooking | Percent of Households | | | | | |
| | Breakfast | | Lunch | | Dinner | |
	Iloilo City	San Jose City	Iloilo City	San Jose City	Iloilo City	San Jose City
Oil-based/ fried	79	70	55	53	71	67
Water-based/ soup	86	77	65	83	78	83
Grilled	1	1	2	3	6	5
Smoked	0	11	0	14	0	3

Source: ADB. 2015. Promoting Sustainable Energy for All in Asia and the Pacific - Energy Access for Urban Poor. TA 8946. *Household survey, 2018.*

The surveys found that cooking sessions in Iloilo City and San Jose City vary from 23 to 34 minutes depending on the meal prepared (whether for breakfast, lunch, or dinner) or an average of 33 minutes per meal. The women of the family are the designated cooks in 82% of the households in San Jose City and 77.6% of households in Iloilo City. Within a day, family member in-charge of cooking spend an average of 1 hour and 39 minutes for preparing three meals. Figure 2 shows the average length of time spent in preparing for each meal per day, by city.

Figure 2: Average Cooking Sessions per Meal Type, Per Day
(minutes)

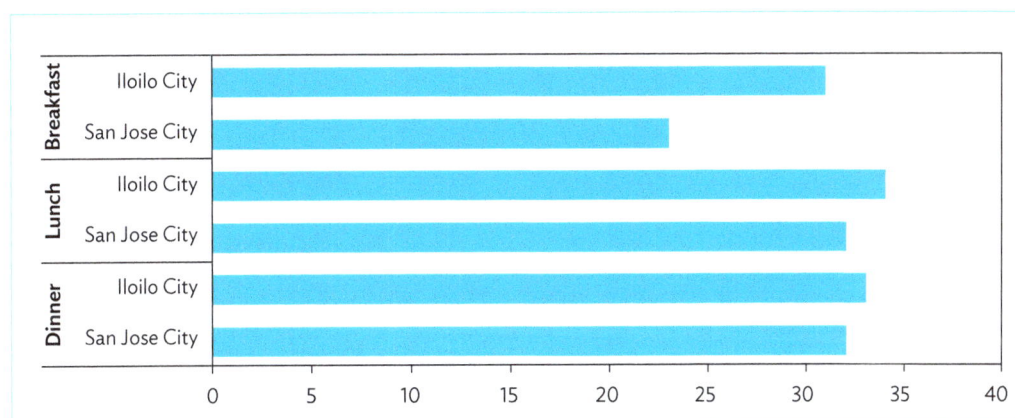

Source: ADB. 2015. Promoting Sustainable Energy for All in Asia and the Pacific - Energy Access for Urban Poor. TA 8946. *Household survey, 2018.*

2.3. Kitchen Ventilation

Kitchen ventilation is an important factor in the level of HAP. Kitchens with better ventilation are those that are typically free from obstruction and have more spaces for air to circulate, or have enough structures such as windows or air vents for dispersing heat and allowing for air exchanges that can disperse pollutants. On the other hand, kitchens with limited ventilation sources are those that are enclosed or have features that limit air circulation and does not allow pollutants to freely disperse into the atmosphere. These, according to WHO, cause greater risk to household members due to accumulation of pollutants from fuel burning and other sources, allergens such as molds, and vectors such as mosquitoes.[17]

The assessment of kitchen layout and ventilation sources shows that majority of kitchens in Iloilo City (68.6%) and San Jose City (65.2%) had either low to non-existent sources of ventilation or were structured in a way that restricted circulation of air in and out of the kitchen. This condition was also evident in the amount of soot accumulation on kitchen walls noted in 57% of the kitchens in San Jose City and 40% of kitchens in Iloilo City. Furthermore, only 1% of households in San Jose City and 7% of households in Iloilo City have exhaust fans to aid ventilation in the kitchen. Box 2 presents details on how household ventilation conditions were evaluated during the field visits to the study sites.

Box 2: Evaluating Household Ventilation, Wind Speed, and Ambient Temperature

During the surveys, kitchen ventilation was assessed and categorized by identifying structural aspects of the respondents' kitchens such as the location of the cooking area in relation to the living area, the number of walls obstructing air flow, and the presence of natural and mechanical ventilation structures such as windows, exhaust vents or exhaust fans in the cooking area. To facilitate the assessment of the impacts of the structural aspects of the kitchens, categories that describe typical kitchen layouts (A to D) were formulated. Category A is assumed to have the most space for air to immediately disperse, while D is the most enclosed, offering the least opportunity for air to circulate. Eventually, these were simplified into two categories: **vented** or those households with relatively better sources of ventilation and with the least physical structures obstructing the flow of air and **unvented** or those with comparably less sources of ventilation and with physical structures that obstruct the efficient flow of air in the kitchen and the rest of the household. Appendix 1 provides details on how kitchens were categorized based on kitchen layout and sources of ventilation.

When air flow is unconfined, the pollutants can be diffused faster into a larger volume of air, preventing emission concentration build-up and household air pollution can be expected to remain at relatively lower levels. To find out the effects of ventilation levels to indoor temperature and wind speed, measurements were taken in all respondents' households for both ventilation levels. The average measurements for both cities are shown in the table below.

Comparisons between ventilation levels show that higher average wind speed were recorded in kitchens with better ventilation levels, giving more chance for emissions from cooking to be dispersed from the kitchen to the surrounding environment. Kitchens with poor ventilation levels, on the other hand, exhibited lower wind speeds, so the opposite scenario can be reasonably expected where pollutants are not as easily dispersed to the surrounding or external environment. On the other hand, no major difference in indoor temperature was noted between the two ventilation levels.

continued on next page

---

[17] World Green Building Council. 2018. Healthier Homes, Healthier Planet Guide. London.

Box 2 continued

Wind Speed and Temperature Per Kitchen Ventilation Type

Study Site	Factors that May Affect Household Air Quality	Vented	Unvented
Iloilo City (highly urbanized city)	Wind speed (m/s)	0.10	0.05
	Temperature (°C)	31.05	31.43
San Jose City (component city)	Wind speed (m/s)	0.16	0.03
	Temperature (°C)	31.34	31.03

Source: ADB. 2015. Promoting Sustainable Energy for All in Asia and the Pacific - Energy Access for Urban Poor. TA 8946. *Household survey, 2018.*

To gain an idea on whether locational factors (coastal versus landlocked city) had an effect on air circulation and temperature within the households, wind speed and temperature measurements in the above table were compared with the average ambient temperature and wind speed in the study areas. Climatological data from the Department of Science and Technology Philippine Atmospheric, Geophysical and Astronomical Services Administration estimated the mean temperature and wind speed for the year the survey was taken (2018) in measurement sites closest to Iloilo City and San Jose City. These data showed very little difference in average mean ambient temperature between Iloilo City (28.6°C) and San Jose City (28.5°C) while there were differences in average wind speed. For the average 2018 data Iloilo City's wind speed was 3.5 m/s or what is considered as a gentle breeze. In San Jose City, average wind speed for the same year was 1.5 m/s which was considered light air, closer to a calm almost still wind condition. In comparison with household measurements where wind speed was almost calm to non-existent even in vented conditions, it can be surmised that location did not affect indoor air circulation and temperature for both cities.

Sources: ADB. 2015. Promoting Sustainable Energy for All in Asia and the Pacific - Energy Access for Urban Poor. TA 8946. *Household survey, 2018*; Government of the Philippines – Department of Science and Technology (DOST); Climatological data provided by the Climatology and Agrometeorology Division of DOST- Philippine Atmospheric, Geophysical and Astronomical Services Administration on 4 April 2020.

While ambient temperature did not have notable differences in vented and unvented conditions as shown in Box 2, the level of ventilation may also have a substantial effect on thermal comfort.[18] Theoretically, because of their structure, households with vented kitchens may experience minimal increases in temperature even while cooking is being done. These were evident in differences in temperature measurements in spaces and on structures like posts or pillars near cookstoves. Figure 3 shows measurements taken in a kitchen that has a relatively better level of ventilation. The thermal reading of the space near the cookstove is 33°C (Figure 3, Example 1), which is close to ambient temperature measurements taken prior to cooking. On the other hand, Example 2 from the same figure shows that structures such as the post, despite being not obstructive in nature, still retained heat from the cookstove, characterized by the 43.8°C temperature recorded by the thermal scan.

From the measurements taken, the ambient temperature in vented and unvented kitchens for both study sites did not show any notable differences. However, it is interesting to note that while thermal readings done in spaces near the cookstoves show an average of 2°C increase from the ambient temperature, measurements on structures like posts or pillars near cookstoves were much higher at 43.8°C or 10.8°C higher than the ambient temperature as recorded by the thermal scan in a vented kitchen. This is shown in Figure 3, Example 2. This basically shows that such structures, despite being non-obstructive in nature, still retained more of the heat from the cookstove. Occupants near or directly in contact with the structure would therefore feel the higher temperature.

[18]   Raish, J. n.d. Thermal Comfort: Designing for People. Edited by W. Land and A. McClain for The University of Texas at Austin, School of Architecture (Center for Sustainable Development).

Figure 3: Temperature and Flow of Heat in a Vented Kitchen

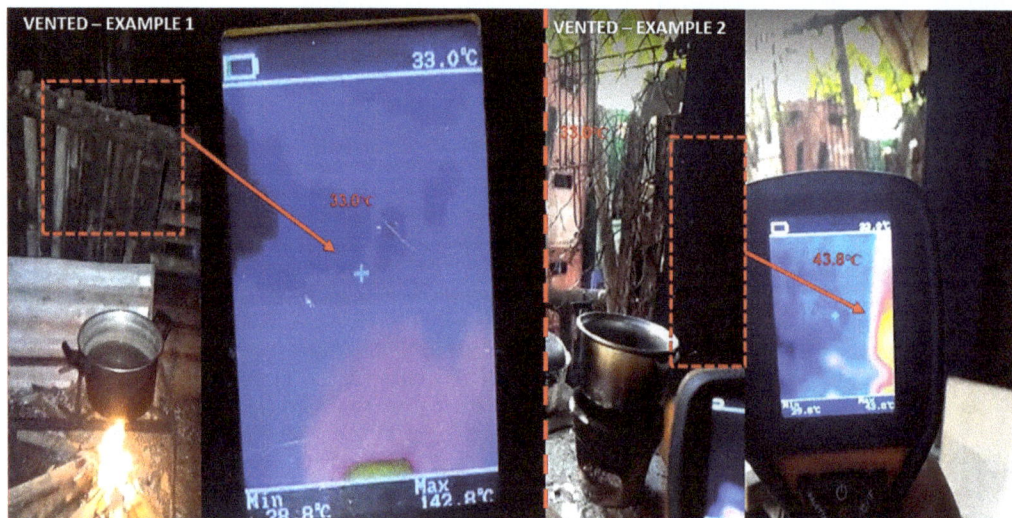

Source: ADB. 2015. Promoting Sustainable Energy for All in Asia and the Pacific – Energy Access for Urban Poor. TA 8946. *Field emission testing. 2018.*

This retention of heat is more prominent in kitchens with low ventilation sources or with plenty of obstructive structures. In Figure 4, Example 1 shows the plume from cooking is about 37.5°C, affecting the room due to the relatively higher temperature of the nearby wall (34.3°C); Example 2 shows a more evident flow and retention of heat. The adjacent wall to the cookstove exhibits a temperature of 44.6°C, confirming that the physical structures near the heat source acts like a sink. In these two examples, the walls that retain heat can later radiate this to the same area, causing higher than normal temperatures and discomfort to the occupants.

Figure 4: Temperature and Flow of Heat in an Unvented Kitchen

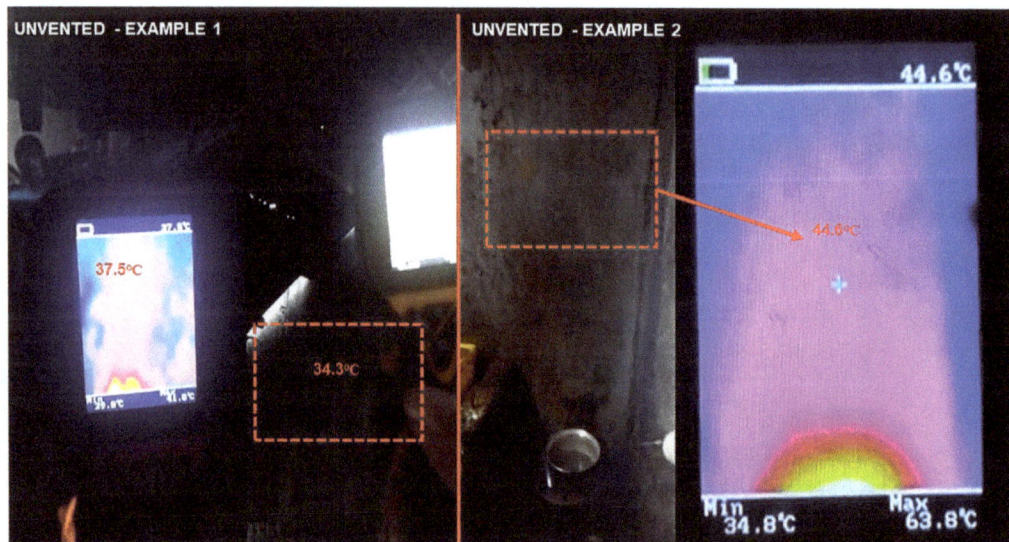

Source: ADB. 2015. Promoting Sustainable Energy for All in Asia and the Pacific – Energy Access for Urban Poor. TA 8946. *Field emission testing. 2018.*

2.4.   Cookstoves and Fuel Preferences

Majority of households in Iloilo City (46.8%) exclusively use traditional cookstoves while only 25.4% exclusively use modern cookstoves. In San Jose City, 27.5% of households exclusively use modern cookstoves, while only 16.0% of households exclusively use traditional cookstoves. This confirms that households in both Iloilo City (25.4%) and San Jose City (55.5%) also practice fuel stacking or use multiple types of fuels and stoves.

Fuel stacking is a common household fuel choice decision practiced globally as a coping strategy to market price fluctuations and unreliable fuel supply availability. It allows households to accommodate different food preparation preferences as well as cook for longer periods as required for specific dishes such as slow-cooked viands or smoked/grilled meats (footnote 13). From a case study by the Food and Agriculture Organization of the United Nations (FAO), charcoal and fuelwood are considered important residential fuels in the Philippines. Fuelwood is readily available and oftentimes gathered for free, while charcoal is used because of cultural preferences for certain types of food. This study also explains that partiality to using traditional cookstoves in the Philippines is not only because of convenience but also due to local traditions and preferences for flavor enhancement brought about by the use of charcoal and/or fuelwood and a common belief that cooking on such cookstoves are more economical especially for slow-cooked meals.[19]

In both study sites, fuel stacking was observed in households—whether they are primarily using traditional, gas, or electric stoves; majority of households in San Jose City notably practice it. Those using gas stoves practice stacking with traditional cookstoves, represented by 13.9% of households in Iloilo City and 36.5% of households in San Jose City (Table 3).

Table 3: Household Cookstove Preference

Cookstove Preference	Iloilo City (%)	San Jose City (%)
Exclusive Traditional	**46.8**	**16.0**
Charcoal	38.3	3.0
Fuelwood	4.0	6.5
Both	4.5	6.5
Exclusive Modern	**25.4**	**27.5**
LPG	18.9	26.0
Butane	5.5	1.0
Electricity	1.0	0.5
Practicing Fuel Stacking	**25.4**	**55.5**
Primary Traditional	9.9	19.0
Secondary Gas	9.4	19.0
Secondary Electric	0.5	0
Primary Gas Stove	13.9	36.5
Secondary Traditional	13.9	36.5
Secondary Electric	0	0
Primary Electric	1.5	0
Secondary Traditional	1.5	0
Secondary Gas	0	0
No Answer/ Erroneous data	**2.5**	**1.0**
TOTAL	**100**	**100**

LPG = liquefied petroleum gas.
Source: ADB. 2015. Promoting Sustainable Energy for All in Asia and the Pacific - Energy Access for Urban Poor. TA 8946. *Household Survey. 2018.*

[19]   Remedio, E.M., 2009. An analysis of sustainable fuelwood and charcoal production systems in the Philippines: A Case Study. Criteria and Indicators for Sustainable Woodfuels: Case Studies from Brazil, Guyana, Nepal, Philippines and Tanzania. Food and Agriculture Organization: Rome. http://www.fao.org/3/i1321e/i1321e00.pdf

Traditional cookstoves can accommodate the use of either charcoal or fuelwood that are easily accessible locally in both cities, or in the case of fuelwood can also be gathered for free from the immediate environment whether as fallen tree stems or wood scraps from abandoned construction materials. Charcoal, on the other hand, can be bought in retail packs at corner stores ranging from 0.150 to 0.500 kilograms in weight or by bulk in sacks. Based on an emissions inventory commissioned by the Iloilo City local government unit (LGU) in 2016, a sack of charcoal is reported to weigh 29 kilograms on average. Charcoals bought by the sack are usually delivered or distributed by major suppliers by bulk of 200–300 sacks to local markets (colloquially referred to as "Super") thrice a week. Fuelwood, if bought, is usually by 2 kilogram bundles.

Box 3: Charcoal and Fuelwood Sources

The Philippines is one of the countries in Asia that still produce fuelwood and charcoal commercially for household cooking. In 2016, it was estimated that majority of the residential sector's energy consumption was for biomass, of which 47.3% was for fuelwood, 11.8% for charcoal, 6.3% for agricultural waste, and the remaining energy consumed are electricity and petroleum products. The Philippines charcoal market is projected to advance further, from its 2019 value of $500 million to $688.2 million in 2030 in a compounded annual growth rate of 2.9% during the 2020–2030 forecast period. Lump charcoal, or charcoal made by slowly burning wood pieces in the absence of oxygen, is the largest category in the Philippine charcoal market due to the cost-effectiveness in its processing.

Philippine forestry statistics data indicate that 266 cubic meters or 26.6% of wood produced from plantations are converted into fuelwood (27 cubic meters [m³] or 2.7%) and charcoal (239 m³ or 23.9%) with the remaining 733 m³ processed into various wood products. Only 11 m³ of the fuelwood makes its way out of the country, with the rest consumed locally.

In Iloilo City, which is known as the Philippines' home of grilled delicacies, charcoal continues to be a popular cooking fuel. Grilling is done whether in big or small food establishments, and charcoal is purchased by a lot of low-income households in small quantities from the corner *sari-sari* (variety or convenience) store. An analysis of the charcoal value chain in Iloilo City estimated that demand as of 2017 required clearing 1.36 hectares of wood land per year to supply the city's needs. Iloilo City itself does not produce charcoal rather it is supplied by producers from 13 municipalities in Iloilo Province and five municipalities in nearby Guimaras Province. Guimaras Province produces 70%–80% of the charcoal Iloilo City consumes.

Sources: Government of the Philippines – Department of Energy. 2019. Philippine Energy Profile 2017–2040. Manila; PSMarketresearch. 2019. Philippines Charcoal Market Research Report: By Application, Type - Industry Opportunity Analysis and Growth Forecast to 2030. Manila; Government of the Philippines – Department of Environment and Natural Resources. 2018. *Forest Management Bureau - Philippine Forestry Statistics 2018.* Manila; United States Agency for International Development. 2017. *Analysis of the Charcoal Value Chain in Iloilo City (Final Report).* September. Manila. Prepared under the USAID Building Low Emission Alternatives to Develop Economic Resilience and Sustainability (B-LEADERS) Project.

A bundle of fuelwood can last an average of 5 days or 1–7 days in Iloilo City where majority of households primarily use traditional cookstoves and 16 days or 1–90 days in San Jose City where majority of households primarily use LPG stoves. In the same way, packs of charcoal get consumed quicker in Iloilo City, with households indicating that a pack lasts only a day on average while in San Jose City a pack lasts 7 days on average before they purchase another pack. For sacks of charcoal, the averages for the two cities (34 days in Iloilo City and 31 days in San Jose City) are close.

LPG fuel comes in tanks of varying tank sizes, i.e., 2.7 kilograms (kg), 7 kg, or 11 kg. Majority of respondents using LPG stoves indicated that they use 11 kg tanks. A few, however, use the lower-sized LPG tanks. For those using the 11 kg tanks, the average number of days a tank is consumed is the same for both cities, at 71 days, before requiring tank refill (Table 4).

Table 4: Number of Days that Each Fuel Lasts Before Requiring New Purchase

Duration of use (in no. of days)	Fuelwood Bundle (2 kg)	Charcoal Pack (0.150–0.500 kg)	Charcoal Sack (29 kg)	LPG 2.7 kg	LPG 7 kg	LPG 11 kg
ILOILO CITY						
Average	5	1	34	30	30	71
Maximum	7	7	180	…	…	548
Minimum	1	1	3	…	…	15
Mode	1	1	30	…	…	30
SAN JOSE CITY						
Average	16	7	31	28	32	71
Maximum	90	30	90	…	84	672
Minimum	1	1	4	…	14	7
Mode	7	1	15	…	28	56

kg = kilogram, LPG = liquefied petroleum gas, … = not available.
Source: ADB. 2015. Promoting Sustainable Energy for All in Asia and the Pacific - Energy Access for Urban Poor. TA 8946. Household Survey. 2018.

Fuel storage location varies per household. More often, fuel (whether firewood, charcoal, or LPG) are placed indoors or specifically within the kitchen while some households opt to place firewood or charcoal outside, taking measures to ensure these solid fuels are kept dry and away from sources of moisture. In Iloilo City, 38% of respondents indicated that they provided a storage area for their charcoal although these specified storage areas were not necessarily protected against moisture. In San Jose City, respondents indicated that storage areas are usually in any random available space whether inside or outside the house, as long as these kept the fuels away from rain. Otherwise for these solid fuels in both cities, no standards are in place as to level of dryness or quality from point of purchase or acquisition.

3. Field Emissions Tests on Household Cooking Fuel

An *Emission Inventory Study for Iloilo City* was conducted and published in 2015 using 2010 as the base year for the data with some inputs from 2011 and 2012. The study was done in coordination with the Department of Environment and Natural Resources, GIZ, and Clean Air Asia through the Association of Southeast Asian Nations (ASEAN)–German project "Clean Air for Smaller Cities (CASC) in the ASEAN Region" with the aim of aiding the development of an effective air quality management system. The study concluded that among the pollutants included in the inventory, cooking was found to contribute the highest amount of PM_{10} due to the amount of charcoal used by households for cooking.[20] The significant findings of the emissions inventory study highlighted the need for further inspection, through actual field survey, of local kitchens to establish the extent and impact of HAP on indoor air quality to which household occupants are exposed to.

Emission testing of household cookstoves have been conducted by WHO since the early 1990s under controlled settings using the water boiling test (WBT) method or through simulated cooking tests. The WBT method is a standard test developed by the Global Alliance for Clean Cookstoves (GACC) to measure how efficiently a stove uses fuel to heat water in a cooking pot.[21] WHO has acknowledged that the results of these tests did not reflect actual emissions in households during actual home cooking activities, and only provided a limited understanding of factors that drive the variability of emissions over geographic scales.[22] Field emission testing is therefore considered a complementary exercise that can improve the understanding of HAP impact, which is critical to the pursuit of policy and behavioral changes to promote the use of cleaner fuels.

Thus, field emission testing was conducted in conjunction with the household surveys to gather actual field measurements and determine the amount of air pollutants, namely of $PM_{2.5}$, CO, SO_2, and NO_2, emitted by the various fuel–technology combinations employed by households in the study sites, and its effect on HAQ using WBT (Box 4).

The conduct of the field emission testing also allowed the responsible team to document the household respondents' cooking fuel of choice for the conduct of the WBT. Here, it was found that in Iloilo City, 53.7% of households preferred using charcoal, 4.5% of households preferred fuelwood, and 29.8% preferred LPG as cooking fuel. In San Jose City, 21% of households preferred fuelwood, 9% preferred charcoal, and 66% of households indicated a preference for LPG as cooking fuel. For both cities, only a small percentage of households used butane or electricity as shown in Table 5.

---

[20] GIZ. 2015. *ASEAN – German Technical Cooperation, Clean Air for Smaller Cities in the ASEAN Region*. Emission Inventory of Major Air Pollutants in Iloilo City (Final Report). Unpublished.

[21] GACC. 2014. Global Alliance for Clean Cookstoves. The Water Boiling Test Version 4.2.3.: Cookstove Emissions and Efficiency in a Controlled Laboratory Setting.

[22] Edwards, R., et. al. n.d. WHO Indoor Air Quality Guidelines: Household fuel Combustion (Review 2: Emissions of Health-Damaging Pollutants from Household Stoves). WHO.

Box 4: Water Boiling Test Methodology Employed During the Field Emission Testing

For the field emission tests done in the two cities of Iloilo and San Jose, the water boiling test (WBT) required a standard size of cooking pot and a standard volume of water. In this case, for each type of stove, the same cooking pot were used with a fixed 1-liter volume of water. Households were requested to prepare their stoves according to the type of fuel that they primarily use for cooking (either charcoal, fuelwood, liquefied petroleum gas [LPG], butane, or electricity). The initial water and air temperatures are recorded. After this step, the stove is then ignited, if kindling material needs to be used (in the case of traditional stoves), or the stove is simply turned on in the case of LPG, butane, or electric hot plates. The test continues until the water in the pot reaches the boiling point of 100 degrees Celsius.

Several instruments were used to measure emissions inside the kitchen while the water boiling test is being conducted. These instruments were placed near the stove at a distance of around 1 meter or less to simulate the exposure levels to air pollutants of the person preparing the meals. Instruments include sensors for the measurement of levels of identified pollutants throughout the test. These include pollutants which, according to the World Health Organization (WHO), have the most negative effects on health: particulate matter ($PM_{2.5}$), carbon monoxide (CO), sulfur dioxide (SO_2), and nitrogen dioxide (NO_2). $PM_{2.5}$ are fine, inhalable particles that remain suspended in the air compared with heavier particles like dust, often from burning fuels. CO is an odorless, colorless gas which, in households, is produced when cooking especially during incomplete combustion such as when using wood or charcoal stoves. This is dangerous because exposure to CO displaces oxygen in the body and in extreme amount can lead to poisoning. SO_2 is also a colorless gas but has a strong odor; it is more often produced by the burning of fossil fuels such as coal and oil. NO_2 is one of a group of highly reactive gases that gets in the air from the burning of fuel. It can react with other chemicals in the air and form particulate matter and ozone, both of which are harmful to the respiratory system.

Measurements taken included: (i) before the test starts (before cooking), (ii) after the stove is ignited (kindling), (iii) measurements during the actual boiling phase (during cooking), and (iv) after the water boils (after cooking). During cooking, one measurement is taken every 3 minutes until either the test ends (water boils) or until a minimum of three measurements are taken. For modern stoves (LPG, butane, electric), at this same stage one measurement is taken every 2 minutes until either the test ends (water boils) or until a minimum of three measurements have been taken. A hot wire anemometer was also used to measure the wind speed inside the kitchen as an indicator for ventilation. The field WBT was given a hard limit of around 30 minutes even if boiling point is not reached. In this case, the temperature of the water after this hard limit is taken and the test is prematurely concluded.

Sources: ADB. 2015. Promoting Sustainable Energy for All in Asia and the Pacific - Energy Access for Urban Poor. TA 8946. *Field Emission Test.* 2018; US EPA website. *Indoor air quality.* WHO website. *Household air pollution: Pollutants.* Accessed 2 September 2020.

Table 5: Household Fuel Preference During Field Emission Testing Phase

City	Charcoal (%)	Fuelwood (%)	LPG (%)	Butane (%)	Electricity (%)
Iloilo City	53.7	4.5	29.8	9.0	3.0
San Jose City	9.0	21.0	66.0	3.5	0.5

LPG = liquefied petroleum gas.
Source: ADB. 2015. Promoting Sustainable Energy for All in Asia and the Pacific - Energy Access for Urban Poor. TA 8946. *Field Emission Test.* 2018.

3.1.   Linking Cooking Fuel Use to Indoor Air Quality

WHO has devised the *Guidelines for indoor air quality: household fuel combustion*[23] to provide standards that can help define "clean" or standards for clean cooking technologies in the home. Any type of fuel–technology combination is considered "clean" if it meets the emission reduction targets (ERT) set by the guidelines for $PM_{2.5}$ and CO pollutants.[24] The ERT values (Table 6) also specify the level of emissions from these pollutants within which cookstoves and cooking fuels pose minimal health risks, providing a decision point for selecting the less health-destructive device and fuel for household use.

Table 6: World Health Organization Emission Rate Target Recommendations for Household Fuel Combustion

Pollutant	Unvented Kitchen	Vented Kitchen
Particulate matter	0.23 mg/min	0.80 mg/min
Carbon monoxide	0.16 g/min	0.59 g/min

g = grams, mg = milligrams, min = minute.
Source: WHO indoor air quality guidelines: household fuel combustion (Recommendation 1: Emission Rate Targets).

The per-city average measurements of $PM_{2.5}$ and CO emission rates taken from vented and unvented kitchens for each fuel–technology combination during the field emission tests were compared with WHO ERT standards to determine which among the fuel–technology combinations were "cleaner" than the others, and would lead to better HAQ that can provide the least amount of harm to health for household members.

Unsurprisingly traditional cookstoves, whether utilizing charcoal or fuelwood, had emission rates that greatly exceeded ERT levels both for $PM_{2.5}$ and CO in both cities. In Iloilo City, $PM_{2.5}$ emissions of LPG stoves for both vented and unvented kitchens also exceeded the ERT guideline values. For vented kitchens, LPG emissions exceeded the ERT guideline values by 87.5% while emissions in unvented kitchens were 400% higher than the ERT guideline value. In San Jose City's unvented kitchens $PM_{2.5}$ emission from LPG stoves exceeded ERT levels by 38.8% but were within ERT levels in vented kitchens. On the other hand, stoves using butane had mixed results, with emissions in vented kitchens in San Jose City and unvented kitchens in Iloilo City both exceeded WHO ERT values and the reverse is true for the remaining ventilation conditions. Electric stoves meanwhile had $PM_{2.5}$ emission rates in both cities for both vented and unvented kitchens that were within the ERT guideline values (Figure 5).

CO emission rate measurements from both cities for traditional cookstoves using either charcoal or fuelwood were also found to have exceeded ERT guideline values. However, CO emission rates for the other fuel-technology combinations for both cities and in both vented and unvented kitchens were within the ERT guideline values (Figure 6).

[23]   WHO. *WHO Guidelines for indoor air quality: household fuel combustion.*

[24]   UN. 2018. Accelerating SDG 7 Achievement: Policy Briefs in Support of the First SDG 7 Review at the UN High-Level Political Forum. Policy Brief #2: Achieving Universal Access to Clean and Modern Cooking Fuels, Technologies and Services.

**Figure 5: Comparison of Particulate Matter Emission Rates of Vented and Unvented Kitchens
in Iloilo City and San Jose City with the World Health Organization
Emission Reduction Target Values**

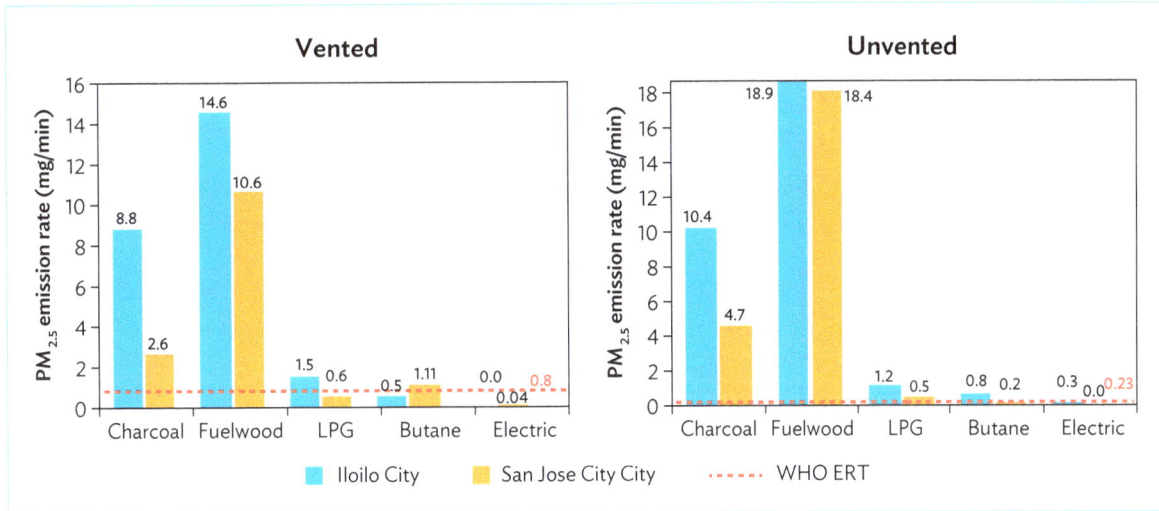

ERT = emission reduction target, LPG = liquefied petroleum gas, mg = milligram, min = minute, PM 2.5 = particulate matter, WHO = World Health Organization.
Source: ADB. 2015. Promoting Sustainable Energy for All in Asia and the Pacific - Energy Access for Urban Poor. TA 8946. *Field Emission Test.* 2018.

**Figure 6: Comparison of Carbon Monoxide Emission Rates of Vented and Unvented Kitchens
in Iloilo City and San Jose City with the World Health Organization
Emission Reduction Target Values**

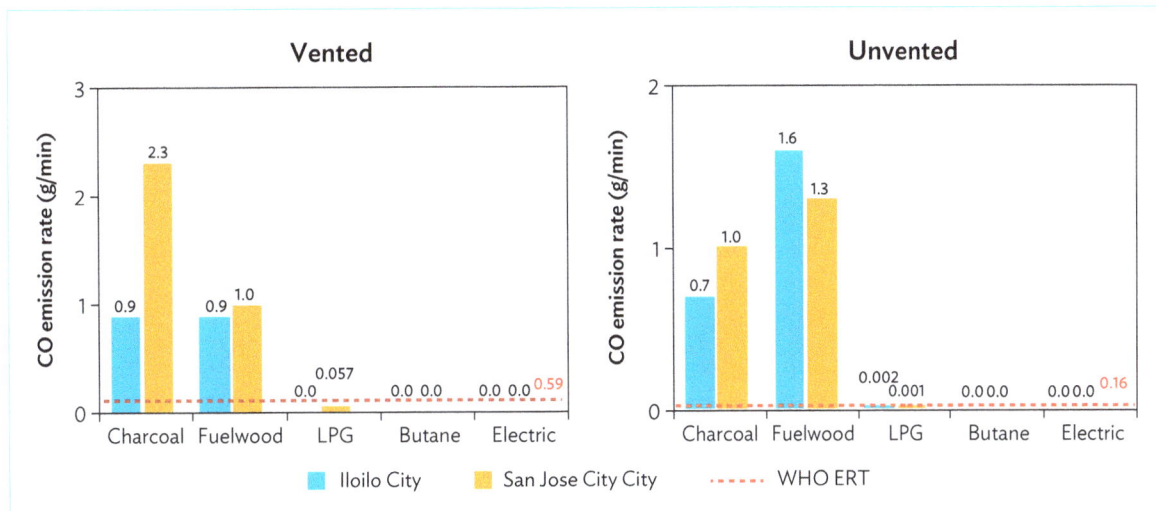

CO = carbon monoxide, ERT = emission reduction target, g = gram, LPG = liquefied petroleum gas, min = minute, WHO = World Health Organization.
Source: ADB. 2015. Promoting Sustainable Energy for All in Asia and the Pacific - Energy Access for Urban Poor. TA 8946. *Field Emission Test.* 2018.

That the amount of $PM_{2.5}$ and CO emitted by cookstoves using traditional fuels such as charcoal and fuelwood is beyond the recommended targets by WHO signifies that continued use of such fuels for cooking can compromise the health of those exposed. It argues strongly for switching toward cooking devices and fuels that emit less amounts of these pollutants such as LPG, butane, and electric-powered stoves.

3.2.   Indoor Air Quality and Health

According to WHO, a kitchen's ventilation status has a direct effect on the level of indoor air pollution and the exposure to emissions of household members.[25] The measurements gathered from the field tests were therefore assessed to find out whether ventilation conditions of kitchens in the study sites also had noticeable direct effects to HAQ. As earlier noted in Chapter 2.3, ventilation and physical (obstructive) structures were shown to have effects on thermal comfort.

The average emission concentrations of each pollutant for each city and for each of the ventilation conditions are presented in Table 7.[26] The results showed that ventilation did affect the emission concentration within the households. The emission concentration measurements for $PM_{2.5}$, SO_2, and NO_2, were, as expected, higher in unvented conditions. However, the measurements of CO were found to be inconsistent, showing higher values in vented kitchens compared to unvented ones. Upon careful examination of emission measurements, these were found to be due to higher CO measurements for charcoal in both cities in vented conditions.

CO is the main emission from burning of charcoal as opposed to $PM_{2.5}$ when burning fuelwood. An article published by US EPA in 2013 discussed how CO is generally not emitted from airtight woodstoves or fireplaces.[27] Since traditional Philippine cookstoves are not airtight, ventilation allows CO to escape in vented kitchens. This could explain why CO is higher in vented kitchens compared to unvented kitchens.

Table 7: Average Emission Concentration of Pollutants from Charcoal and Fuelwood in Vented and Unvented Kitchens During Cooking, for Iloilo City and San Jose City

Pollutant	Iloilo City		San Jose City	
	Vented	**Unvented**	**Vented**	**Unvented**
$PM_{2.5}$ ($\mu g/m^3$)	365.57	369.07	318.57	602.35
CO (mg/m^3)	30.77	28.84	54.38	44.59
SO_2 (mg/m^3)	0.11	0.75	3.66	4.25
NO_2 (mg/m^3)	0	0.002	0.20	0.35

CO = carbon monoxide, mg = milligrams, m^3 = cubic meter, NO_2 = nitrogen dioxide, $PM_{2.5}$ = particulate matter, SO_2 = sulfur dioxide, μg = microgram.
Source: ADB. 2015. Promoting Sustainable Energy for All in Asia and the Pacific - Energy Access for Urban Poor. TA 8946. *Field Emission Test.* 2018.

---

[25]   WHO. 2016. *Burning opportunity: clean household energy for health, sustainable development, and wellbeing of women and children.* Geneva.

[26]   The average emission concentration measurements during cooking of $PM_{2.5}$, CO, SO_2 and NO_2 from traditional cookstoves using fuelwood and charcoal were used because emissions from the modern fuel–technology combinations were below the detection limits of the measurement instruments employed during the field emission tests

[27]   Oanh, Nguyen Thi. 2012. *Integrated air quality management: Asian Case Studies.* Edited by Kim Oanh N. T. New York: CRC Press.

The impact of indoor air pollution on health can be assessed using the US EPA AQI. The index that provided comparisons between ranges of emissions and its possible consequences to health were formulated with the general purpose of making information about air quality accessible and easily understandable. It provided ranges of emission concentrations with equivalent levels of health concern for each of the four pollutants. Categories range from the least health affective (good), followed by the categories moderate, unhealthy for sensitive groups, unhealthy, very unhealth until the last category which is the most health-destructive (hazardous). The US EPA symbolized each of these levels with specific colors (Table 8).[28]

Table 8: United States Environmental Protection Agency Air Quality Index for Particulate Matter, Carbon Monoxide, Sulfur Dioxide, and Nitrogen Dioxide

Level of Health Concern	Symbol/Color for Each Level of Health Concern	Range of Emissions for Each Pollutant			
		$PM_{2.5}$ ($\mu g/m^3$)	CO (mg/m^3)	SO_2 (mg/m^3)	NO_2 (mg/m^3)
Good	Green	0-12	0-5.5	0-0.10	-
Moderate	Yellow	12.1-35.4	5.6-11.7	0.11-0.41	-
Unhealthy for Sensitive Groups	Orange	35.5-55.4	11.8-15.5	0.42-0.64	-
Unhealthy	Red	55.5-150.4	15.6-19.2	0.65-0.87	-
Very Unhealthy	Purple	150.5-250.4	19.3-38.0	0.88-1.73	1.3-2.5
Hazardous	Maroon	>250.4	>38.0	>1.73	>2.5

CO = carbon monoxide, mg = milligrams, m^3 = cubic meter, NO_2 = nitrogen dioxide, $PM_{2.5}$ = particulate matter, SO_2 = sulfur dioxide, μg = microgram.
Source: United States Environmental Protection Agency. 2012. *Revised air quality standards for particles pollution and updates to the air quality index (AQI)*; US EPA. n.d. *AirNow Factsheet: Air Quality Forecasts and Observations.*

While the US EPA AQI ranges were calculated based on ambient air quality, the household measurements' equivalent health effects are expected to be similar. Ranges of pollutants within good and moderate categories provide little to no negative health effects to household members. People who are active outdoors, and members of the "sensitive group (e.g., people with asthma, the elderly, and children) become vulnerable to the effects of pollutants once these breaches the moderate range. However, for pollutants reaching the unhealthy to hazardous levels, all who are exposed may begin experiencing adverse health effects with those from the sensitive group susceptible to greater risk and more adverse health effects. Locally, the Philippine Clean Air Act of 1999 or Republic Act No. 8749 also produced national ambient air quality guideline values (NAAQGV) which is more or less patterned from the US EPA AQI but is more conservative in its emission range per health concern category.[29]

During the field emission testing phase, emission concentration measurements were taken at different stages: before cooking, kindling, during cooking, and after cooking. Because ventilation was seen to affect emission concentration, the observations for vented and unvented kitchen conditions were separated. The comparison of the resulting average emission concentration measurements for vented and unvented kitchens with the US EPA AQI for each city for both $PM_{2.5}$ and CO are presented graphically in Figure 7 and Figure 8.

---

[28] US EPA. 2014. Air Quality Index - A guide to air quality and your health. February. https://www.airnow.gov/sites/default/files/2018-04/aqi_brochure_02_14_0.pdf

[29] Government of the Philippines, Department of Environment and Natural. *2015. Environmental Management Bureau: National Air Quality Status* Report *2008–2015.*

Figure 7: Particulate Matter Emission Contribution to Indoor Air Quality, at Various Points of the Cooking Process
($\mu g/m^3$)

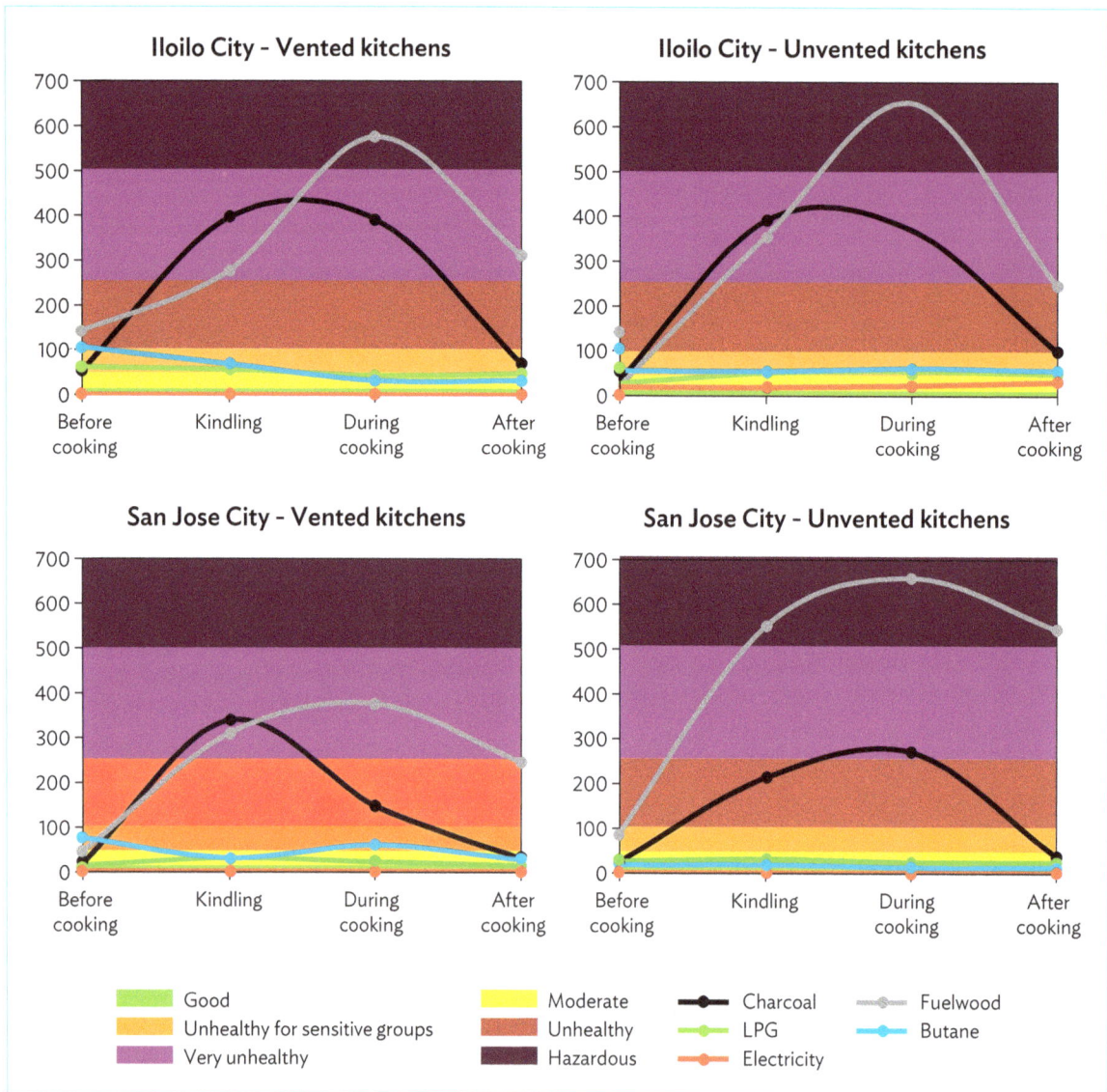

LPG = liquefied petroleum gas, m^3 = cubic meter, μg = microgram.
Source: ADB. 2015. Promoting Sustainable Energy for All in Asia and the Pacific - Energy Access for Urban Poor. TA 8946. *Field Emission Test*. 2018.

- **Before cooking.** Even prior to cooking, $PM_{2.5}$ emission concentrations were recorded in kitchens in both Iloilo City and San Jose City. This is to record traces of $PM_{2.5}$ in the ambient environment left behind from previous cooking activities that were not completely circulated clean out of the kitchen or easily disturbed by ventilation.

- **Kindling stage.** Kindling, for traditional cookstoves, involve use of dry, thin scraps of wood layered under the charcoal or fuelwood and/or twisted paper as igniting agent to facilitate kindling. Sometimes, this will also entail fanning the flame to ensure that the charcoal or fuelwood are sufficiently burning and produce enough heat for cooking. As seen in Figure 7, high levels of $PM_{2.5}$ were observed for charcoal and fuelwood during this stage regardless of ventilation conditions for both cities. For butane and LPG stoves in Iloilo City, $PM_{2.5}$ breached the "moderate" range and was recorded at the lower limits of the "unhealthy for sensitive groups" range while these remained within moderate AQI levels in San Jose City.

- **During cooking.** $PM_{2.5}$ emissions were highest during cooking for both cities, also regardless of the ventilation stage for traditional cookstoves using charcoal or fuelwood. The concentration of $PM_{2.5}$ were especially higher for fuelwood, reaching hazardous levels for both ventilation conditions in Iloilo City and for unvented conditions in San Jose City. In Iloilo City, LPG and butane emission concentrations became lower during this stage at kitchens with better ventilation conditions and remained stable in kitchens with poor ventilation conditions.

- **After cooking.** Even after cooking, $PM_{2.5}$ levels from traditional cookstoves using fuelwood were still measured at concentrations that were considered very unhealthy in vented kitchens in Iloilo City and at the upper limit of the "unhealthy" range for kitchens with poor ventilation conditions. $PM_{2.5}$ from charcoal became lower but still at levels considered unhealthy for sensitive groups. Emissions from LPG also remained at levels considered unhealthy for sensitive groups for both ventilation conditions. In San Jose City, $PM_{2.5}$ emissions from traditional cookstoves using fuelwood decreased but still at unhealthy levels in kitchens with better ventilation conditions but remained at hazardous levels in kitchens with poor ventilation conditions. Emission measurements from charcoal, butane, and LPG stoves all decreased to levels, which were considered moderate to good.

Figure 8: Carbon Monoxide Emission Contribution to Indoor Air Quality, at Various Points of the Cooking Process
(mg/m³)

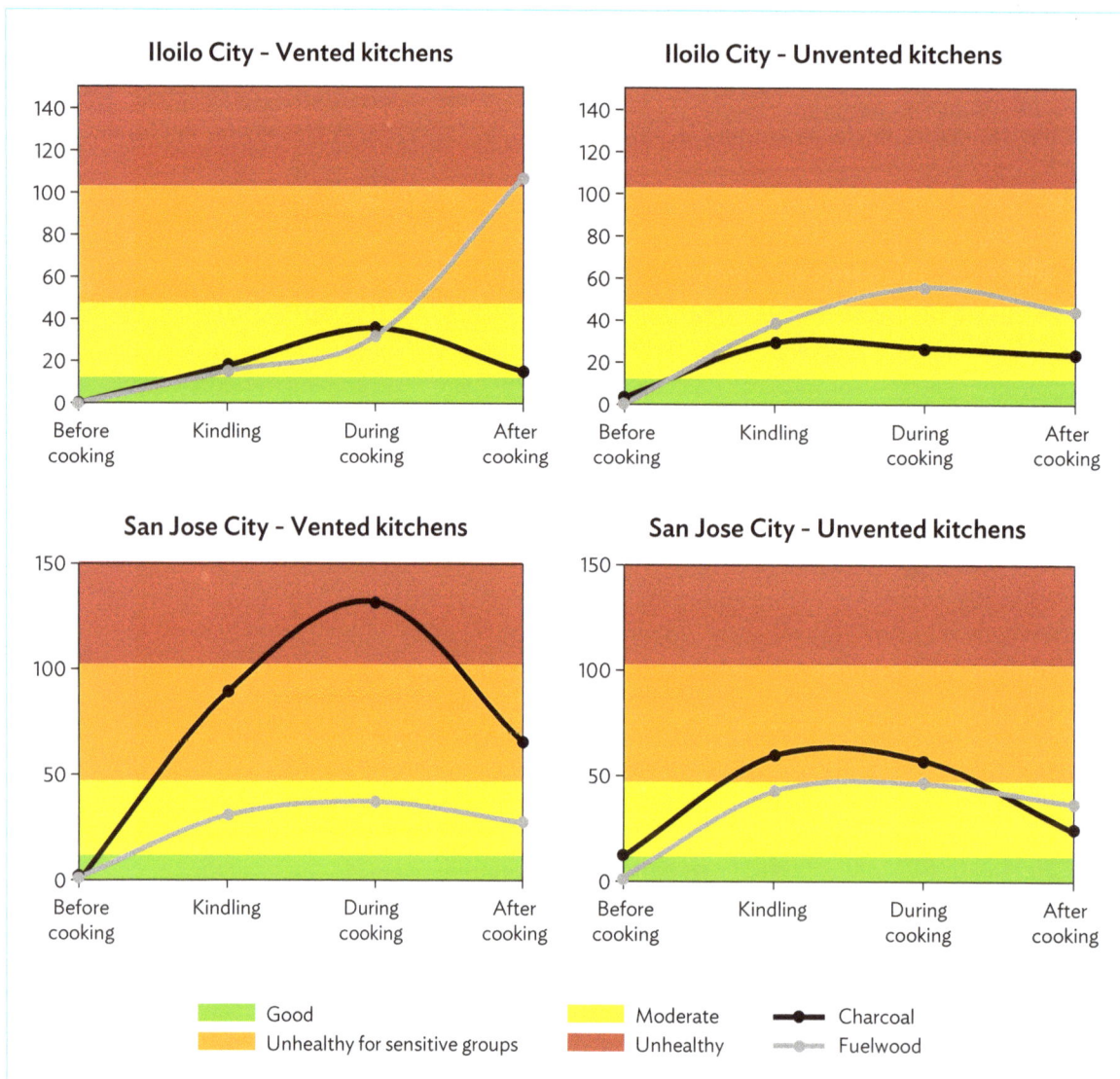

mg = milligram, m³ = cubic meter.
Source: ADB. 2015. Promoting Sustainable Energy for All in Asia and the Pacific – Energy Access for Urban Poor. TA 8946. *Field Emission Test.* 2018.

- **Before cooking.** In Iloilo City, CO measurements in vented and unvented kitchens using fuelwood before cooking were undetected while very small amounts of CO were detected in households using charcoal. In San Jose City, measurements taken from vented and unvented kitchens using both charcoal and fuelwood were also either very small or undetected, except for measurements from unvented kitchens that used charcoal, where the amount of CO emission almost went over the upper limit of the AQI range still considered good for the exposed members of the households.

- **Kindling stage.** As was explained earlier, the kindling stage for fuelwood and charcoal entails a longer process for igniting flames. In Iloilo City's vented kitchens, CO emissions from both fuelwood and charcoal stoves increased but remained within moderate AQI levels. However, for unvented kitchens, CO emissions from fuelwood increased to within the range considered unhealthy for sensitive groups while emission from charcoal increased but stayed within moderate AQI levels.

In San Jose City's vented kitchens, CO measurements from fuelwood were within the moderate AQI range while CO emissions from charcoal increased to levels considered unhealthy for all exposed household members. In the unvented kitchens, CO emissions from both fuelwood and charcoal increased to levels considered unhealthy for all household members exposed.

- **During cooking.** CO emissions further increased during cooking for all kitchen ventilation conditions in both cities. In Iloilo City, households using fuelwood recorded CO emission levels that were moderate (for ventilated kitchens) or at the lower limit of the level that AQI considers as unhealthy for all members of the household (for unvented kitchens). CO emission measurements for households using charcoal for both ventilation conditions also increased from that of the kindling stage but were still within moderate AQI levels.

  In San Jose City, CO emissions in both vented and unvented kitchens using fuelwood also increased and reached levels considered unhealthy for sensitive groups. For households using charcoal, CO measurements in vented kitchens remained within unhealthy levels while in unvented kitchens measurements slightly decreased from during the kindling stage but were still at a level considered unhealthy for all exposed members of the households.

- **After cooking.** CO emission levels further increased in vented kitchens in Iloilo City for kitchens using fuelwood where it was considered unhealthy for all members of the households while measurements from unvented kitchens slightly decreased but were still considered unhealthy for sensitive groups. For households using charcoal, CO emissions from vented and unvented kitchens both decreased but remained within the moderate AQI range.

  A possible explanation as to why CO levels further increased in Iloilo City is that in traditional cookstoves, one cannot just turn off the stove at the end of cooking, therefore the remaining embers still emit CO even after the water has boiled during the WBT.

  In San Jose City, CO emissions after cooking using fuelwood and charcoal decreased for both vented and unvented kitchens. For vented kitchens, CO measurements for kitchens using fuelwood fell to within moderate AQI levels while in unvented kitchens the measurements remained unhealthy for sensitive groups. For households using charcoal, in vented kitchens, CO decreased but was still within the range considered unhealthy for all exposed household members. For unvented kitchens, CO emissions decreased to levels considered just moderately harmful to the exposed members of the household.

CO measurements for LPG, butane, and electric stoves were very low, or at times zero for all stages of cooking and for both ventilation conditions in both cities. Summarizing the results from both cities and comparisons with both WHO ERT standards and US EPA AQI, electric stoves is seen to have the least negative impact on indoor air quality and health. Traditional fuels, charcoal and fuelwood proved to be the most precarious to the household members' health. These fuel types have shown considerably high levels of $PM_{2.5}$, CO, and SO_2, which can significantly worsen household air quality while emissions from NO_2 were undetectable by the measuring devices used during the field emission tests.

4. Controlling External Factors in Laboratory Tests

Aside from the WBT conducted during the field emission testing phase, WBT was also conducted in a laboratory setting. As prescribed by the GACC, this involved a simplified approximate simulation of cooking conditions at an environment that is controlled for ventilation, kitchen dimension, and other possible externalities in order to derive results that can be comparable across the fuel–technology combinations being tested. The laboratory test strictly followed the comprehensive WBT protocol by GACC and assesses stove performance or how efficiently the stoves tested used fuel to heat water in a cooking pot, as well as the emissions produced during cooking (footnote 21). The results of the laboratory tests provided additional inputs for determining which of the technologies is the cleaner option and can be used as a guide to any decision for shifting to cleaner cooking alternatives. These complements the results of the field emission tests, which showed the actual emission conditions at the household level.

Four fuel–technology combinations were tested in the laboratory. Traditional cookstoves, as well as fuelwood and charcoal were brought back from the two cities. For modern cookstoves a Super Kalan or a 2.5 kg LPG tank with an attached cookstove and a 1,200 watt single-coil electric cookstove were used. On the other hand, since no ICS were observed as being used in both cities during the household surveys, data on Philippines-developed ICS from a workshop laboratory test done by StovePlus Academy in Iloilo City[30] were used for comparison with the results of the laboratory tests that were conducted for the four fuel–technology combinations.

The laboratory WBT consisted of three phases that immediately followed each other:

- **Cold-start high-power phase** – the laboratory test begins at room temperature and uses a pre-weighted fuel prepared beforehand and 1.5 liters of water in room temperature to boil.

- **Hot-start high-power phase** – water is replaced immediately after it reaches boiling point and this second phase starts immediately, a new set of pre-weighted fuel is used.

- **Simmer phase** – the simmer phase is conducted by allowing water to simmer for 45 minutes just below boiling point to simulate long cooking processes. This allows for the measurement of fuel required to perform such tasks.

---

[30] The 2017 StovePlus Academy Workshop in Iloilo City was conducted by the Group for the Environment, Renewable Energy and Solidarity (GERES) which is a nonprofit nongovernment organization, and co-sponsored by ADB under its Energy for All Initiatives.

4.1.   Emission Concentration of Fuels

The laboratory WBT–emission testing allowed for measuring $PM_{2.5}$, CO, SO_2, and NO_2 concentrations that leave the stove. Because of its controlled nature, it is used in comparing performances of different fuel–technology combinations from each other and not the extent of human exposure to emissions or indoor air pollution concentration. In this regard, the GACC deemed field testing as an essential step to justify claims about actual impacts on emissions, and the resulting effects specifically to HAP and more broadly to overall air pollution resulting from the stoves (footnote 21).

The results of the laboratory emission tests showed that traditional stoves utilizing charcoal recorded $PM_{2.5}$ emission concentrations that were 5,000 times higher than that emitted by an LPG stove. Fuelwood emitted an even higher concentration of $PM_{2.5}$, which was 95,000 times higher compared with that from the LPG stove (Table 9). This complements the results of the field emission testing wherein charcoal and fuelwood were also determined to have the highest $PM_{2.5}$ emissions.

CO emissions for electric and LPG stoves were recorded as "lower than the detection limit" of the measuring devices used. The use of electricity does not involve combustion; therefore, CO is not produced. On the other hand, LPG stoves may produce little to no CO depending on the condition of the stove. Otherwise, LPG undergoes complete combustion which produces only CO_2. This is why CO emissions were recorded as "lower than the detection limit" of the measuring devices used. In contrast cookstoves using charcoal and fuelwood emitted large amounts of CO. Since a lot of the organic components present in both charcoal and fuelwood undergo incomplete combustion, copious amount of CO is produced when these fuels are burned. Upon exposure, CO can enter the blood stream through the lungs and eventually affect how oxygen reaches the body's organs and tissues. Very high, unhealthy or even hazardous levels of CO can affect people with cardiovascular diseases and infants the most.

The SO_2 and NO_2 emissions of both LPG and electricity were also recorded as "lower than the detection limit" of the two methods used. In contrast, charcoal and fuelwood emitted detectable concentrations of these gases. Comparing the two, fuelwood emitted significantly higher amounts of SO_2. This indicates that fuelwood contains more sulfur, which reacts with oxygen upon combustion. NO_2 measurements between charcoal and fuelwood are relatively close in value, with charcoal higher by a small margin.

Table 9: Average Emission Concentration of Each Pollutant by Fuel Type

Fuel Type	$PM_{2.5}$ ($\mu g/m3$)	CO (mg/m3)	SO_2 (mg/m3)	NO_2 (mg/m3)
Charcoal	157,051.7	390.3	1.1	0.61
Fuelwood	2,794,694.9	224.6	19.6	0.57
Liquefied Petroleum Gas	29.4	none	<Detection Limit	<Detection Limit
Electricity	10.9	none	<Detection Limit	<Detection Limit

CO = carbon monoxide, mg = milligram, m^3 = cubic meter, NO_2 = nitrogen dioxide, $PM_{2.5}$ = atmospheric Particulate Matter that have a diameter of less than 2.5 micrometers SO_2 = sulfur dioxide, μg = microgram.
Source: ADB. 2015. Promoting Sustainable Energy for All in Asia and the Pacific - Energy Access for Urban Poor. TA 8946. *Laboratory testing data*. 2018.

Considering the results of the field emission tests and as confirmed by the laboratory test, traditional cookstoves utilizing charcoal or fuelwood release air pollutants at levels that are found to be "unclean" by WBT standards, and compromise air quality and the health of household members at all stages of cooking. Going by the survey results, this means that 46.8% of the households in Iloilo City and 16% of the households in San Jose City exclusively using traditional fuel–technology combinations expose themselves to poor HAQ due to "unclean" cooking practices. This is also true for the 25.4% of households in Iloilo City and 55.5% of households in San Jose City practicing fuel stacking. This exposure greatly affects the health of the members of these households, especially those who are frequently tasked with preparing the meals.

Similarly, the laboratory emission measurements also affirm the field emission test results' comparison with WHO ERT wherein in both cities, traditional cookstoves utilizing charcoal or fuelwood cannot be considered as "clean" cooking fuel–technology combinations and continued use of these technologies present significant health risks to household members due to the level of emission concentration that these fuel–technology combinations produce.

4.2.   Cookstove and Cooking Fuel Efficiencies

The laboratory test process allowed for the assessment of how well the fuel–technology combinations performed as evidenced by the resulting thermal efficiency ratings. Thermal efficiency is the ratio of the heat that was utilized for cooking to the total heat produced by the combustion of the fuel. A higher thermal efficiency rate indicates a greater ability to transfer the heat produced by the combustion of the fuel into the pot utilized for cooking. This in turn translates to reduced cooking time and less amount of fuel utilized. Table 10 summarizes the results of the tests and includes the results of the ICS tests done during the ADB-supported activity, the 2017 StovePlus Academy Workshop in Iloilo City, for comparison purposes (footnote 30).

Table 10: Thermal Efficiencies of the Different Cookstoves and Cooking Fuel

Stove	Traditional[a]		Improved (Local)[b]			Modern[a]	
	Cement Stove	Cement Stove	Biolexis (Gasifier stove)	Wonder Kalan	Mabaga Kalan (Rocket stove)	Gas Stove[c]	Electric (Coil) Stove[c]
Cooking Fuel	Fuelwood	Charcoal	Rice Hull	Charcoal	Charcoal	LPG	Electric
Thermal Efficiency (%)	10.4	5.2	16	13	27	26.5	33.4
Total Operating Time (minutes)	43.3	43.3	32	28.2	28.2	23.4	18

LPG = liquefied petroleum gas.
[a] Source of data for traditional and modern stoves ADB. 2015. Promoting Sustainable Energy for All in Asia and the Pacific - Energy Access for Urban Poor. TA 8946. *Laboratory testing data.* 2018.
[b] Source of data: Stove+ Academy, 2017 (According to the report of this workshop, a single water boiling test was done as an exercise for the participants, most of which having no prior experience doing the test. The single water boiling test was also not done in a controlled environment or a laboratory. The results should be viewed in this context.)
[c] The laboratory test used the Super Kalan gas stove, which is attached to a 2.7 kilogram LPG tank. For electric stoves, the laboratory test used a 1,200 watt single-coil electric stove.

The thermal efficiency of traditional cookstoves ranked lowest among the stoves tested, at 10.4% fuel efficiency for the cookstove utilizing fuelwood and a lower 5.2% for the cookstove utilizing charcoal. The electric stove had the highest thermal efficiency at 33.4% followed by the LPG stove with 26.5% efficiency.[31] Comparing with the results from the StovePlus Academy workshop measuring efficiencies of ICS developed and produced in the Philippines, the Mabaga Kalan (rocket stove) which utilized charcoal measured a thermal efficiency level which was slightly higher than that of the LPG stove while the Biolexis (gasifier stove) utilizing rice hull as fuel, and the Wonder Kalan utilizing charcoal had thermal efficiencies which were just slightly higher than the results for the test done on traditional cookstove utilizing fuelwood. It should be noted, however, that WBT results from the workshop were not from a completely controlled environment and done only as an exercise by workshop participants in a university laboratory upon the direction of a university professor. Nevertheless, the results somehow show that, with their higher thermal efficiencies there is reason to further study the potential of ICS for household transitioning to cleaner cooking technologies and fuels. Considering fuel stacking by households, the use of ICS can provide better cookstove efficiency resulting in lower fuel cost and emissions.

Testing for thermal efficiency provided approximate values as to the time (in minutes)[32] and amount of fuel that were used up in heating a specific quantity of water for each of the fuel–technology combinations tested. Table 11 shows that traditional cookstoves utilizing charcoal consume the most amount of fuel in order to boil 1.5 liters of water (953 grams [g]) followed by fuelwood and LPG stoves. The electric stove meanwhile consumes 0.21 kilowatt-hour (kWh) of electricity in the 18 minutes that it takes to put the 1.5 liters of water to a boil.

**Table 11: Laboratory Results – Time and Amount of Fuel or Electricity
Needed to Boil 1.5 Liters of Water**

Stove	Traditional		Modern	
	Cement Stove	Cement Stove	Gas Stove[a]	Electric (Coil) Stove[b]
Cooking Fuel	Fuelwood	Charcoal	LPG	Electric
Unit of fuel used to boil 1.5 liters of water (grams)	716 g	953 g	91.5 g	0.21 kWh
Minutes used to boil 1.5 liters of water	43	43	23	18

g = grams, kWh = kilowatt-hour, LPG = liquefied petroleum gas.
Note: (a) The Super Kalan stove with 2.7 kilogram LPG tank was used for the laboratory tests; (b) A 1,200-watt single coil electric stove was used for the laboratory tests
Source: ADB. 2015. Promoting Sustainable Energy for All in Asia and the Pacific - Energy Access for Urban Poor. TA 8946. *Laboratory testing data.* 2018.

4.3.   Estimating Stove and Fuel Costs

To estimate stove and fuel costs, additional information gathered during the survey on the average time consumed in preparing meals for each city, fuel consumption behavior such as mode of purchase (whether via retail or per sack), cost of fuel, amount consumed, and prevailing local market prices of various cooking technologies available in the two cities were used.

---

[31]   The type of LPG stove used may have contributed to the lower than expected level of efficiency in this case since only a Super Kalan, an LPG stove variant with a 2.7 kg LPG tank attached directly to a single stove, was used for the laboratory testing that may have affected the expected result of higher thermal efficiency for LPG when compared with electric stoves.

[32]   This is referred in the WBT as the temperature-corrected specific fuel consumption values (specifically used for charcoal, fuelwood, and LPG), which is a measure of the amount of fuel that was consumed in boiling 1.5 liters of water.

Prevailing Local (Market) Prices of Cooking Fuel

The summary of the average prices of cooking fuels for each city gathered during the household survey phase of the study is shown in Table 12.

Table 12: Price of Cooking Fuel in Study Sites
(₱)

| Study Site | Fuelwood | Charcoal | | LPG | | |
	Bundle (2 kg)	Pack (0.325 kg)	Sack (29 kg)	2.7 kg	7 kg	11 kg
San Jose City	0.71 ($0.01)	9.08 ($0.17)	171 ($3.21)	265 ($4.98)	308 ($5.79)	642 ($12.07)
Iloilo City	4.25 ($0.08)	9.96 ($0.19)	419 ($7.88)	240 ($4.51)	660 ($12.41)	866 ($16.28)

kg = kilogram, LPG = liquefied petroleum gas.
Note: Conversion of US dollar to Philippine peso is $1.00 = ₱53.1953 (average of October–December 2018 exchange rates, taken from http://www.bsp.gov.ph/statistics/spei_new/tab12_pus.htm.
Source: ADB. 2015. Promoting Sustainable Energy for All in Asia and the Pacific - Energy Access for Urban Poor. TA 8946. *Household Survey*. 2018.

Fuelwood is the cheapest among all the four cooking fuels (fuelwood, charcoal, LPG, electricity). Households can purchase it by bundles weighing 2 kilograms averaging around ₱ 0.71 in San Jose City and ₱ 4.25 in Iloilo City. It should be noted that this average may have been influenced by the fact that fuelwood can also be gathered for free from the immediate surroundings. The discrepancy in price can be attributed to San Jose City remaining a peri-urban or a predominantly agricultural area where there is an abundance of free fuelwood from branches and twigs from nearby trees compared to the more urban environs of Iloilo City.

Charcoal, on the other hand can be bought either by packs of 0.325 kilograms or by bulk in sacks weighing 29 kilograms. In San Jose City, purchasing by pack costs ₱9.08 ($0.17), and by sack ₱171.00 ($3.21). In Iloilo City, a pack of charcoal costs not too far off at ₱9.96 ($0.19) while a sack costs a lot more at ₱419.00 ($7.88) each. Refilling of LPG tanks, which are distinguished by tank sizes ranging from 2.7 kg, 7 kg, and 11 kg can be done at a cost of ₱265.00 ($4.98) for a 2.7 kg tank, ₱308.00 ($5.79) for a 7 kg tank or ₱642.00 ($12.07) for an 11 kg tank in San Jose City. In Iloilo City, the same tank sizes may cost households ₱240.00 ($4.51) for a 2.7 kg tank, ₱660.00 ($12.41) for a 7 kg tank or ₱866.00 ($16.28) for an 11 kg tank to refill.

Using the above values gathered from the household surveys as summarized in Table 12, the cost per unit of cooking fuel (in grams and kilowatt-hour [kWh]) was computed and summarized in Table 13. The difference in LPG prices between the two cities is due to differences in the transportation costs of supplying the area.[33] Considering its geographical location, it is more costly to ship LPG from Manila to Panay Island where Iloilo is located as compared to transporting it by land to San Jose City, Nueva Ecija. In terms of electricity, the cost of electricity per kWh is also slightly higher

---

[33]   Ludovice, H.V. 2018. "Overview: Philippine Downstream Oil Industry" Presentation for the Department of Energy in the Energy Consumers and Stakeholders Conference with the theme: "E-Power Mo! Communicating Efficiency Across the Energy Sector." 24 April. Hotel Supreme, Baguio City, Philippines.

in Iloilo City than in San Jose City. According to a comparison done in 2018, Iloilo City residents pay the highest electricity rate among urban dwellers in the Philippines, including Metro Manila.[34]

Table 13: Cost Per Unit of Cooking Fuel
(₱)

Stove	Traditional		Modern	
	Cement Stove	Cement Stove	Gas Stove	Electric (Coil) Stove
Cooking Fuel	Fuelwood (g)	Charcoal (g)	LPG (g)	Electric (kWh)
San Jose City	0.0279	0.0004	0.0479	11.12
Iloilo City	0.0307	0.0021	0.0646	12.13

g = gram, kWh = kilowatt-hour, LPG = liquefied petroleum gas, .
Source: ADB. 2015. Promoting Sustainable Energy for All in Asia and the Pacific – Energy Access for Urban Poor. TA 8946. *Household Survey*. 2018.

Up-front cost of Stoves

Aside from fuel costs, the prevailing up-front costs of purchasing each type of cooking technology were also gathered and compared. This illustrates how much a household has to spend to acquire a new cooking technology.

The up-front costs of cement cookstoves are ₱91.29 ($1.72) in Iloilo City and ₱234.15 ($4.40) in San Jose City. This difference is owed to the fact that such cookstoves have variable or no standardized designs with no actual product regulatory standards to adhere to. In Iloilo City, these cement stoves generally have common uncomplicated physical characteristics while those in San Jose City, while following a similar round shape, come in varying sizes and have modifications for ease in placing the cooking fuel, all of which contribute to this price difference.

Electric single-coil stoves are moderately priced for both cities, costing ₱983.33 ($18.49) in Iloilo City and ₱1,225.00 ($23.03) in San Jose City. Single and double-burner LPG stoves had the highest up-front cost because the price includes the cost of purchasing a filled LPG tank and its various accessories such as the valve regulator, hose, and hose clamp. The cost of LPG stoves is also variable, depending on the number of burners. The average cost of a Super Kalan, which uses a 2.5 kg LPG tank is ₱1,940.00 ($36.47) in Iloilo City and ₱1,965.00 ($36.94) in San Jose City. The average price of a single-burner LPG cookstove with an 11 kg LPG tank is ₱4,295.75 ($80.75) in Iloilo City and ₱4,071.75 ($76.54) in San Jose City. Lastly, the average price of purchasing a double-burner LPG cookstove with an 11 kg LPG tank is ₱5,445.86 ($102.37) in Iloilo City and ₱4,967.00 ($93.37) in San Jose City. Table 14 summarizes comparative up-front costs of traditional, improved, and modern cookstoves in each city. ICSs were not observed to be used by households during the survey and field emission testing phase. The up-front cost for four ICS available in the Philippine market were nevertheless included in the summary for comparison purposes.

---

[34] *Manila Standard*. 2018. Power rates in Iloilo highest among urban centers. 18 November.

Table 14: Up-front cost of Traditional, Improved, and Modern Cookstoves

Stove	Traditional[a]		Improved (Local)[b]				Modern[a]			
	Cement Stove	Cement Stove	Biolexis (Gasifier Stove)	Papa-Brik Stove	Wonder Kalan	Mabaga Kalan	Super Kalan[c]	Single Burner Gas Stove[d]	Double Burner Gas Stove[d]	Electric (Coil) Stove
Cooking Fuel	Fuelwood	Charcoal	Rice Hull	Pili nut shells	Charcoal	Charcoal	LPG	LPG	LPG	Electric
Iloilo City	₱91.29 ($1.72)		₱2,659.77 ($50.00)	₱797.93 ($15.00)	₱372.37 ($7.00)	₱531.95 ($10.00)	₱1,940.00 ($36.47)	₱4,295.75 ($80.75)	₱5,445.86 ($102.37)	₱983.33 ($18.49)
San Jose City	₱234.15 ($4.40)						₱1,965.00 ($36.94)	₱4,071.75 ($76.54)	₱4,967.00 ($93.37)	₱1,225.00 ($23.03)

LPG = liquefied petroleum gas.
Note: conversion of US dollar to Philippine peso is $1.00 = ₱53.1953 (average of October–December 2018 exchange rates, taken from http://www.bsp.gov.ph/statistics/spei_new/tab12_pus.htm).
[a] Calculated using data from the household survey, including market survey in each of the cities during the field visits for the price of the various cooking technologies and apparatuses.
[b] StovePlus Academy, 2017 Workshop Completion Report.
[c] The up-front cost of a Super Kalan consists of the cost of the stove and the cost of a filled 2.7 kilogram LPG tank
[d] The up-front cost of a single and double burner gas stove consists of the cost of the stove, a filled 11 kilogram LPG tank, and the attachments from stove to tank such as the valve regulator, hose, and hose clamp.

Average Amount of Fuel Used by a Household Per Year for Each City

Using survey data from Chapter 2 on the time spent by households in preparing meals for each city and data on fuel use in Table 11, the average annual amount of fuel used by a household for each of the four main fuel–technology combination were derived. Results of these are shown in Table 15 below. The computations show that if households were to use traditional cookstoves exclusively, they could consume between 565.63 kg and 600.53 kg of fuelwood or between 424.99 kg and 451.21 kg of charcoal annually. On the other hand, exclusive use of LPG could lead to annual consumption of 100.47 kg to 106.67 kg of LPG, while exclusive use of a single electric (coil) stove could lead to consumption of 298.05 kWh to 316.44 kWh of electricity annually.

Table 15: Average Amount of Fuel Used by a Household Per Year for Each City

Stove	Traditional		Modern	
	Cement Stove	Cement Stove	Gas Stove	Electric (Coil) Stove
Cooking Fuel	Fuelwood (kg)	Charcoal (kg)	LPG (kg)	Electric (kWh)
San Jose City	600.53	451.21	106.67	316.44
Iloilo City	565.63	424.99	100.47	298.05

LPG = liquefied petroleum gas.
Source: Computed from results of *Household Survey* and *Laboratory WBT*, ADB. 2015. Promoting Sustainable Energy for All in Asia and the Pacific - Energy Access for Urban Poor. TA 8946.

Annual Cost to Households of Different Fuel–Technology Combinations

To approximate the annual cost to household for each of the four fuel-technology combinations as summarized in Table 16, the time and amount of fuel or electricity needed by each of the four fuel-technology combinations to boil 1.5 liters of water were used (Table 11). These values, together with the cost per unit of cooking fuel (Table 13) were then computed against the annual average time spent by households in preparing meals for each of the two cities to derive the annual costs..

Table 16: Annual Cost to Households of Using Fuel to Cook Water-Based Food

Stove	Traditional		Modern	
	Cement Stove	Cement Stove	Gas Stove	Electric (Coil) Stove
Cooking Fuel	Fuelwood	Charcoal	LPG	Electric
Iloilo City	₱958.83 ($18.02)	₱18,414.01 ($346.16)	₱6,893.73 ($129.59)	₱3,838.44 ($72.16)
San Jose City	₱151.78 ($2.85)	₱15,800.10 ($297.02)	₱4,813.62 ($90.49)	₱3,314.35 ($62.31)

LPG = liquefied petroleum gas.
Note: Computed from results of *Household Survey* and *Laboratory WBT*, ADB. 2015. Promoting Sustainable Energy for All in Asia and the Pacific – Energy Access for Urban Poor. TA 8946; conversion of Philippine peso to US dollar is ₱53.1953: $1.00 (average of Oct – Dec 2018 exchange rates taken from http://www.bsp.gov.ph/statistics/spei_new/tab12_pus.htm).

Surprisingly and contrary to what households commonly believe as more economical especially for slow cooking, using charcoal turned out as the most expensive among the different cooking fuels. In Iloilo City, where there is more demand for charcoal, the 53.7% of respondents who are primarily using charcoal for their cookstoves, could shell out as much as ₱18,414.01 ($346.16) for charcoal per year, though they may not feel the burden because purchases are done in small amounts, on a per pack basis. For households in San Jose City, the 9% of households who use charcoal for cooking can spend around ₱15,800.10 ($ 297.02) annually.

This is followed by LPG stoves where households exclusively using a single-burner stove with an 11 kg LPG tank will spend around ₱6,893.73 ($129.59) in Iloilo City and ₱4,813.62 ($90.49) in San Jose City annually. Households who will exclusively use a single-coil electric stove similar to the one used during the laboratory test will incur an additional annual electricity cost of ₱3,838.44 ($72.16) in Iloilo City or ₱3,314.35 ($62.31) in San Jose City annually. This levels off to an additional to the monthly electricity bill of around ₱319.87 ($6.01) in Iloilo City and ₱276.20 ($5.19) in San Jose City. As may be expected the least-expensive cooking fuel is fuelwood, with households who use this exclusively only spending around ₱958.83 ($18.02) in Iloilo City and ₱151.78 ($2.85) in San Jose City annually.

5. The Outlook on Shifting to Modern Technology

Switching from traditional fuel–technology combinations to modern ones provides many advantages, not exclusive to households. The benefits include household cost reduction, an improvement in HAQ by reducing emissions of air pollutants that negatively affect the health of household members, contributing to the improvement of ambient air quality, and reducing pollutants' contribution to climate change. Furthermore, switching not only will help achieve the SDG 7 goals but also contribute to achieving other interlinked SDGs.

5.1. Emission and Cost Reduction Outlook with Shifting

From the point of view of HAQ, a switch from traditional fuel-technologies to modern cookstoves can greatly reduce household pollution, especially $PM_{2.5}$ emissions and its attendant health hazards. In terms of reducing greenhouse gas (GHG) emission, the IEA foresees switching to LPG instead of electric stoves will lead to a slight increase in CO_2 but that this will be more than offset by the expected reduction in emissions from ceasing the use of fuelwood or charcoal as cooking fuel.[35] These are compelling reasons for advancing the case for household switching from traditional cooking technologies to cleaner cooking solutions.

Emission Outlook of Shifting

Based on the field emission tests, in San Jose City a switch from cookstove using charcoal can reduce $PM_{2.5}$ concentration to at least 60.04% (charcoal to butane) to as much as 99.32% (traditional charcoal to electric stove). In Iloilo City, switching fuels from charcoal to LPG can lead to reduction in $PM_{2.5}$ emission concentration from 83.36% (charcoal to butane) to as much as 96.55% (traditional fuelwood to electric stove). The possible reductions in emission concentrations from the different exchanges from traditional to modern fuel–technology combinations are presented in Table 17.

---

[35] IEA. 2019. SDG7: Data and Projections (Access to affordable, reliable, and sustainable energy for all). Flagship report. November.

Table 17: Percent of Particulate Matter Emission Concentration Reductions for Different Fuel–Technology Switching Combinations

Traditional Cookstove Fuel	Modern Cookstove Fuel	Iloilo City		San Jose City	
		Vented (%)	Unvented (%)	Vented (%)	Unvented (%)
Charcoal	LPG	89.33	85.04	86.09	91.20
	Butane	92.19	83.36	60.04	95.47
	Electric	No available field samples	93.92	99.32	No available field samples
Fuelwood	LPG	92.75	91.52	94.56	96.40
	Butane	94.69	90.57	84.38	98.15
	Electric	No available field samples	96.55	99.74	No available field samples

LPG = liquefied petroleum gas.
Notes: Computed using average emission concentrations measured in vented and unvented kitchens for both study sites during cooking; For the "no available field samples" entries, this means there were no households surveyed under this category.
Source: ADB. 2015. Promoting Sustainable Energy for All in Asia and the Pacific - Energy Access for Urban Poor. TA 8946. *Field Emission Testing.* 2018.

However, households practicing fuel stacking will not be able to maximize these exchanges in contrast with households switching from exclusively using traditional fuel–technology combinations.

In estimating possible GHG reductions several assumptions were taken into consideration. It was assumed that (i) GHG emissions for electricity considered a 9% transmission and distribution loss; (ii) the GHG computations are approximations as heat exchanges (from one fuel to another) is not the same; and, (iii) values for the exchanges used the number of households in both Iloilo City and San Jose City that were exclusively using traditional and modern cookstoves for the exchanges. Based on these assumptions, two scenarios were computed for, as follows:

- Scenario 1: The minimum values are based on the scenario that wood processed into charcoal or used as fuelwood are harvested from tree plantations and no deforestation occurred. In this case, GHG emissions from using charcoal and fuelwood for cooking can be considered biogenic, that is burning these fuelwood and charcoal returns to the atmosphere the carbon that they absorbed. This means that they are carbon-neutral, net emissions are zero, and they do not contribute to global warming.

- Scenario 2: The maximum values are based on the extreme opposite—that wood stock used as fuelwood and processed into charcoal came from illegal wood cuttings and GHG emissions are therefore non-biogenic and use of such would produce GHG emissions.[36]

A third scenario, where a percentage of charcoal and fuelwood production came from legitimate production chains, is more likely. However, the unavailability of data and percentages did not allow for the quantification of this scenario.

---

[36]   IEA. 2020. Fossil vs biogenic CO2 emissions. IEA Bioenergy Technology Collaboration Programme.

Based on the computations, the potential exchanges of GHG emission reductions from switching are summarized in Table 18.

Table 18: Range of Greenhouse Gas Exchanges from Shifting to Cleaner Fuels
(tCO$_2$e/year)

	Minimum		Maximum	
	LPG	**Electricity**	**LPG**	**Electricity**
Fuelwood	7.0	4.3	(14.6)	(17.3)
Charcoal	26.6	16.3	(90.5)	(100.7)
LPG	0	(11.5)	0	(13.1)
Electricity	0.4	0	0.4	0

( ) = negative, LPG = liquefied petroleum gas, tCO$_2$e = tons of carbon dioxide equivalent.
Sources: (i) The greenhouse gas computations are approximations, as heat of one fuel is not exactly the same as another fuel's; (ii) The electricity emission factor used is from the Luzon-Visayas grid (https://www.doe.gov.ph/electric-power/2015-2017-national-grid-emission-factor-ngef?ckattempt=1); (iii) net calorific value and emission factor data of fuels used in the computation are from Garg, A., K. Kazunari, and T. Pulles. 2006. IPCC Guidelines for National Greenhouse Gas Inventories (Volume 2: Energy).

Cost Outlook of Shifting

The up-front cost of the different fuel–technology combinations in Table 14 and the summary of the annual cost to households of using fuel to cook water-based foods in Table 16 were used in generating estimates of the annual cost to households for shifting from one traditional to modern fuel–technology combinations which is summarized in Table 19.

Due to the very low annual cost of both traditional cookstoves and fuelwood, households switching from these can expect to incur additional annual costs ranging from ₱2,919.43 ($54.88) for shifts to a single-coil electric stove and as much as ₱6,148.93 ($115.59) for shifts to a double burner gas stove with an 11 kg LPG tank.

On the other hand, the steep cost of charcoal as a cooking fuel, especially if households use these exclusively as was shown in Table 16, households switching from these can expect to secure savings ranging from ₱10,991.98 ($206.63) for shifts to a single burner gas stove with an 11 kg LPG tank to as much as ₱14,535.75 ($273.25) for shifts to a single-coil electric stove. Since most households used gas stoves with 11 kg LPG tank, the computations used the price of the 11 kg tank and the equivalent cost of refilling this. Further, the summary of additional costs or savings to households for each traditional to modern fuel–technology shifts included costs for the Super Kalan, which can be used only as an outdoor cookstove because it does not have the safety features present in other LPG tank variants.

**Table 19: Annual Cost of Shifting Exclusively from Traditional Cookstove
to Modern Cookstove**

(₱)

	Iloilo City	San Jose City
Cost of shifting from fuelwood to Super Kalan with 2.7kg LPG tank	3,717.89 $69.89	4,568.28 $85.88
Cost of shifting from fuelwood to Single Burner stove with 11-kg LPG tank	6,072.26 $114.15	4,656.33 $87.53
Cost of shifting from fuelwood to Double Burner stove with 11-kg LPG tank	6,148.93 $115.59	4,716.02 $88.65
Cost of shifting from fuelwood to Single Coil electric stove	2,919.43 $54.88	3,091.75 $58.12
Cost of shifting from charcoal to Super Kalan with 2.7kg LPG tank	(13,737.29) ($258.24)	(11,080.03) ($208.29)
Cost of shifting from charcoal to Single Burner stove with 11-kg LPG tank	(11,382.92) ($213.98)	(10,991.98) ($206.63)
Cost of shifting from charcoal to Double Burner stove with 11-kg LPG tank	(11,306.25) ($212.54)	(10,932.30) ($205.51)
Cost of shifting from charcoal to Single Coil electric stove	(14,535.75) ($273.25)	(12,556.56) ($236.05)

( ) = negative, LPG = liquefied petroleum gas.
Note: Conversion of Philippine peso to US dollar is Php53.1953: $1.00 (average of Oct – Dec 2018 exchange rates taken from http://www.bsp.gov.ph/statistics/spei_new/tab12_pus.htm).
Source: Computed using up-front and annual cost of cookstoves and cooking fuel and is adjusted to take into consideration the additional cost incurred due to cookstove replacement as they reach the end of their lifespan.

This means that households shifting from charcoal can expect to recover the cost they spent for a Super Kalan with a 2.7 kg LPG tank within 2 months, a single-burner stove with an 11 kg LPG tank within four months, a double-burner stove with an 11 kg LPG tank within 5 months or a single coil electric stove from less than a month to a little over a month. The payback period for the various technologies is summarized in Table 20.

Table 20: Payback Period of Equipment from Fuel Savings

Cookstove-Fuel Combinations that Led to Savings	Iloilo City			San Jose City		
	Cost of stove	Average Annual Savings from Shifting	Payback period (in months)	Cost of stove	Average Annual Savings from Shifting	Payback period (in months)
Shifting from charcoal to Super Kalan with 2.7kg LPG tank	₱1,700.00 ($31.96)	₱13,737.29 ($258.24)	1.49	₱1,700.00 ($31.96)	₱11,080.03 ($208.29)	1.84
Shifting from charcoal to Single Burner stove with 11-kg LPG tank	₱3,429.75 ($64.47)	₱11,382.92 ($213.98)	3.62	₱3,429.75 ($64.47)	₱10,991.98 ($206.63)	3.74
Shifting from charcoal to Double Burner stove with 11-kg LPG tank	₱4,579.86 ($86.10)	₱11,306.25 ($212.54)	4.86	₱4,325.00 ($81.30)	₱10,932.30 ($205.51)	4.75
Shifting from charcoal to Single Coil electric stove	₱983.33 ($18.49)	₱14,535.75 ($273.25)	0.81	₱1,225.00 ($23.03)	₱12,556.56 ($236.05)	1.17

LPG = liquefied petroleum gas.
Note: Conversion of Philippine peso to US dollar is ₱53.1953: $1.00 (average of Oct – Dec 2018 exchange rates taken from http://www.bsp.gov.ph/statistics/spei_new/tab12_pus.htm).
Source: Computed using up-frontcost of cookstoves (without fuel) and average annual savings incurred by households for switching. Only switches that resulted in savings were included.

5.2.   Barriers to Shifting to Modern Cooking Technologies

There appears to be a positive view in terms of willingness to shift from traditional to modern cookstoves among survey respondents. Of the households exclusively using traditional cookstoves, 54% in San Jose City and 73% in Iloilo City indicated their willingness to shift to clean cooking. However, it was observed that many of the households are already practicing fuel stacking and enjoying the convenience of both traditional and modern fuel–technology combinations in their households. For example, there are households who use electric rice cookers in conjunction with traditional cookstoves and those who have LPG stoves have a stand-by traditional cookstove for slow cooking or for meals that may be enhanced by the distinct flavor imparted by charcoal or fuelwood when using the traditional cooking method.

Table 21: Households Willing to Shift to Modern Fuels

	Households exclusively using traditional fuels	Households willing to shift
San Jose City (Component City)	32 households	17 households (54%)
Iloilo City (Highly urbanized city)	94 households	69 households (73%)

Source: ADB. 2015. Promoting Sustainable Energy for All in Asia and the Pacific - Energy Access for Urban Poor. TA 8946. *Household survey.* 2018.

Thus, convincing households is not as straightforward as showing them the results of the field emissions tests and laboratory tests and expecting to effect and sustain such a shift. Barriers to shifting to clean cookstoves and fuels do exist and have to be understood. The UN, through its regular policy brief status reports on SDG 7, have identified three major challenges that act as barriers to households' shifting to clean cooking and challenge the efforts to achieve universal access to clean cooking by 2030. These key challenges were grouped into three major factors: the issues of supply, demand, and enabling environment.

The challenge with supply refers to issues regarding the availability of clean technologies. Supply issues relate to the lack of clean, affordable and available alternatives as well as stable supply of affordable clean fuels. Demand barriers, on the other hand, include the cost of fuel and/or device, cultural factors, consumer preferences and behavioral practices such as preference to the taste of food prepared using fuelwood or charcoal, a lack of awareness of the negative impacts of traditional fuel–technology combinations, and the practice of fuel stacking.

The challenge regarding enabling environment refers to the prevailing monetary and fiscal policies that may restrict and inhibit growth, non-prioritization of access to clean cooking, and poor coordination among the various sectors which prevent the promotion and sustained adoption of modern fuel–technology combinations. This third factor was emphasized in the latest UN Policy Brief on advancing SDG 7.[37] Governments are enjoined to enact policies that enable sectoral growth to provide clean and efficient cooking technologies that ensure health, climate and gender impacts. Moreover, international policymakers and donors could support this process through capacity-building of government officials and providing resources to both public and private cookstove and fuels stakeholders.

---

[37]   United Nations Economic and Social Commission for Asia and the Pacific. 2020. Accelerating SDG 7 Achievement in the time of COVID-19: Policy Briefs in Support of the High Level Political Forum of 2020 (Policy Brief: Advancing SDG 7 in Asia and the Pacific).

Among the barriers to switching identified by the UN, the most prominent as cited by the respondents during the household surveys were barriers related to demand. The demand barriers identified by respondents can be further classified under two major categories: (i) the up-front, associated, and perceived costs of shifting to modern cooking technologies; and, (ii) the perceived advantages of maintaining or using traditional cookstoves.

The supply or accessibility is not considered as one of the major barriers for not maximizing or for not exclusively using modern cookstoves in their homes. Modern cookstoves are available in local markets at the two study cities. A significant percentage of households also have indicated that they are aware of the availability of these modern cookstoves in their locality. In Iloilo City, these are represented by 86.6% of households while in San Jose City by 58.5% of households. It should be noted, however, that supply issue would be valid if shifting would be from traditional cookstoves to ICS.

Up-front, Associated, and Perceived Costs of Shifting to Modern Cooking Technologies

Respondents have expressed a reluctance to shifting or exclusively using modern cooking technologies mainly because of the costs associated with it. Of the survey respondents, 54.3% of households in Iloilo City and 42.9% of households in San Jose City indicated their concern about cost. First is the up-front cost of purchasing electric or gas stoves, as well as the necessary accessories to gas stoves such as the LPG tank, valve regulator, hose, and hose clamp.

Second is the perceived additional cost for fuel purchase or additional amount to their electricity bill, which respondents are either reluctant or unable to factor within their limited finances in contrast to the perception that traditional fuels are a lot cheaper than using LPG or electric stoves. These were identified as barriers by 43.6% of households in Iloilo City and 57.1% of households in San Jose City.

However, the additional expense relative to fuel purchase is actually only true for households exclusively using fuelwood with their traditional cookstoves. Households exclusively using charcoal stoves can actually expect to realize a substantial amount of annual savings with a recovery period of about a month for shifting from charcoal to single coil electric stove or up to 5 months for shifting to double burner LPG with 11 kg tank. In the long run if their cashflow allows or if financing support is made available to surmount the up-front costs associated with using modern cookstoves a shift would prove beneficial.

Perceived Advantages of Maintaining or Using Traditional Cookstoves

Aside from barriers related to costs, there is the perception barrier. Households believe that exclusively using traditional cookstoves has its advantages. First, 46.8% of households in Iloilo City and 20.0% of households in San Jose City indicated that using traditional cookstoves are more convenient for them. Secondly, according to 25.5% in Iloilo City and 17.1% in San Jose City, it is faster to cook with traditional cookstoves. Finally, owing to the familiarity of having used such cookstoves over generations, households feel that traditional cookstoves are much safer than switching to modern and unfamiliar cookstoves. In their perception, such stoves (particularly LPG stoves) are susceptible to accidental explosions. This perception is manifested more in Iloilo City (40.4%) than in San Jose City (8.6%) where a larger percentage of households already use gas stoves, although a significant number still used these stoves in combination with traditional cookstoves of their choice. Further to this is the Filipino taste preference and specialized cooking needs that explain the continued use of firewood or charcoal, and often influence the practice of fuel stacking in households.

The Need for an Enabling Environment

On the policy front, there is a dearth of government support and regulation for the clean cooking sector. The government's energy access strategy has been dominated by efforts to meet universal electrification targets, which meant that clean cooking has benefited from far less strategic, institutional and policy-related support than the electricity sector.[38] The Philippine Energy Plan 2016–2030 presents a sectoral road map covering the collective efforts of the energy industry in exploring and harnessing all available energy sources in the country.

Promoting energy access is the sector's contribution to poverty alleviation and people empowerment. Thus, the energy sector is concentrating on implementing its Electrification Roadmap to ensure 100% household access to electricity services by 2020.[39] While the plan includes the development of alternative fuels, the alternative energy agenda in the national energy plan is focused on compressed natural gas (CNG), LPG for auto-LPG program and electric vehicles; all for the transport sector, and none to address the slow progress in the adoption of clean cooking technologies.

This policy gap, if not plugged, will only allow the slow progress in the adoption of clean cooking by households in the Philippines to persist. Thus, a specific road map for access to clean cooking is also necessary and its integration into the national energy plan would help in the pursuit of reducing household air pollution and improving national health. Such a plan may be cascaded to the local governments to guide and encourage LGUs to create awareness and to set local goals and targets toward increasing household access to clean cooking alternatives. Several countries within the region have introduced initiatives to foster access to clean cooking. The Philippines may reference these for examples of policies and program initiatives that may be applicable to overcome similar barriers in the country (Table 22).

Underlying in these government initiatives to increasing access to clean cooking is the need to create a baseline of the population that will be targeted by any intervention, as well as the need for financial subsidy for the poorest of the poor who cannot afford the shift to cleaner cooking technologies. The example in Nepal and Bangladesh shows the importance of initial stocktaking prior to the development of market strategies and institutional strengthening and capacity building. Furthermore, the issue of sustainability of any switch needs to be considered for any policy or a series of policies to create lasting impact. Such are the obstacles that the Philippines also need to hurdle in order to attain universal access to clean cooking.

---

[38]   SEforAll Energizing Finance Report Series: Taking the Pulse 2019. Taking the Pulse of Energy Access in the Philippines.

[39]   Government of the Philippines, Department of Energy. 2016. Philippine Energy Plan 2016–2030.

Table 22: Clean Cooking Initiative Examples from Other Developing Member Countries

Country (Year/s)	Project/ Policy	Details
Bangladesh (2013)	Improved Cookstoves (ICS) program by the Bangladesh's Infrastructure Development Company Limited (IDCOL) in partnership with the World Bank	To increase adoption of clean cookstoves, the program included awareness creation, inclusion of women in the supply chain for income generation and poverty reduction, and the establishment of a results-based framework that incentivized partner organizations in developing ICSs locally that suits customer preference. The program has already resulted to over 20% particulate matter ($PM_{2.5}$) reduction and 90% carbon monoxide (CO) reduction and is further estimated to reach 5 million users by 2021 and reduce firewood use by 58%.
Lao People's Democratic Republic (Lao PDR) (2018)	Lao PDR Clean Cookstove Initiative	The Lao PDR Clean Cookstove Initiative is a partnership between the government and a private sector company. It aims to distribute 50,000 highly efficient cookstoves "SuperClean Stoves" or a forced-air gasifier cookstove, to replace charcoal and wood-burning cooking fires in three targeted provinces: Vientiane, Savannakhet, and Champasack. These cookstoves are ICS certified in reducing household air pollution, use cheap biomass pellets from crop waste such as coffee bean, rice husks, corn cobs, or sawdust, and has fuel chambers of different sizes that allows for heat adjustment during cooking. Use of these ICS reduces carbon dioxide (CO2) and particulate emission by 99% over the previously used open fires, burns efficiently and cuts fuel use by 90%.
Nepal (2015)	Efficient, Clean Cooking and Heating (ECCH) Program	A collaboration between the Energy Sector Management Assistance Program (ESMAP) and the World Bank, the program uses scenarios created by ESMAP for the years 2017–2030 for each region and each type of cookstove currently employed and the type of alternative cookstove to switch to. This aids the country in achieving Clean Cooking Solutions for All (CCS4All) that targets to ensure that all households in Nepal cook with improved cookstoves that will alleviate the adverse indoor air pollution-related health impacts of inefficient cooking alternatives.
Viet Nam (2009)	Quality and Safety Enhancement of Agricultural Production and Biogas Development	This project with ADB increased local nongovernment organization efforts in providing rural livestock farmers with biogas digesters that help in safely disposing animal waste while providing households with modern fuel for cooking. ADB assistance aims to increase this access to 16 of the 63 provinces of Viet Nam.

Sources: World Bank. 2018. Bangladesh: Healthier Homes through Improved Cookstoves. *Results Briefs*. 1 November; World Bank. 2013. Market Acceleration of Advanced Clean Cookstoves in the Greater Mekong Sub-region; World Bank. 2014. CiDev.2019. ERPA signed for Lao PDR Clean Cookstove Initiative. 1 August; World Bank. 2018. Super-clean cookstove, innovative financing in Lao PDR project promise results for women and climate. Feature Story. 23 April; World Bank. 2017. Fact Sheet: Efficient, Clean Cooking and Heating Program. An ESMAP/Sustainable Energy for All Partnership. March. Washington, DC; ESMAP. 2017. Nepal| Fostering Healthy Households through Improved Stoves. 29 March.; SNV. nd. Market Acceleration of Advanced Clean Cook Stoves in the Greater Mekong Sub-Region: End User Adoption Study (Vietnam); ADB. 2013. Energy for All: Increasing Access to Clean Cooking (Brochure).

6. Conclusion

The Philippines is among the 20 countries with the largest population still without access to clean cooking. It is also one of the countries in Southeast Asia with little improvement in the drive to attain universal access to clean cooking by 2030. Because of this, it is an ideal starting point for conducting a localized examination of variables affecting the drive to promote and sustain the use of more efficient and environment-friendly cooking alternatives.

The study revealed country-specific information on the household air quality as well as the issues and barriers hindering access to clean cooking technologies. The field emission tests were able to determine the impact of traditional cooking technologies and fuels on HAP and helped to define what is "clean" for health with respect to the cookstoves commonly employed by Philippine households. The US EPA AQI standard was used to gauge the level of adverse health effects brought about by the various fuel-technology combinations not only to those tasked with preparing most if not all meals, but to all the other members of the households as well.

Aside from complementing and affirming the results of the field emission tests in terms of the intensity of emissions from traditional cookstoves utilizing fuelwood and charcoal, the laboratory tests allowed for the comparative measures of thermal efficiency of the different fuel technologies. The various variables identified through the laboratory tests, combined with information gathered during the household survey and the field emission tests were integral in identifying other factors to be taken into consideration when switching from traditional to the modern cookstoves.

With a better understanding of the problems and issues in the Philippines, there is now more capacity to develop local solutions that addresses specific concerns. As not all countries are the same, globally determined issues such as the UN-identified barriers are useful, but the achievement of SDG 7 clean energy goals can be fast-tracked when solutions adapted for each country are emphasized and implemented.

6.1. Summary Of Study Results

Prevailing Fuel and Technology Preference

- **A significant percentage of households still prefer using traditional cookstoves and fuel stacking is a common practice.** In Iloilo City 46.8% exclusively use traditional cookstoves while another 25.4% use traditional cookstoves alternately with a modern cookstove. In San Jose City, 16% of households use traditional cookstoves exclusively, while 55.5% use traditional cookstoves alternately with modern cookstoves.

- **Kitchen ventilation and structures in the kitchen affect emission concentrations of air pollutants and retention of heat from cooking.** The field emission tests showed that a kitchen's ventilation affects the level of emissions. Ventilated kitchens register lower emissions of 44% for NO_2, 30% for $PM_{2.5}$, and 24% for SO_2 when compared with measurements from unvented kitchens. The presence of structures, like posts, in the kitchen also affected ventilation, preventing the effective dispersal of heat from cooking.

Fuel Effect on Indoor Air Quality

- **For both cities, cookstoves utilizing either fuelwood or charcoal had $PM_{2.5}$ and CO emission rates that greatly exceeded WHO ERT recommendations.** The results of the field emission tests showed that WHO ERT guideline values for $PM_{2.5}$ and CO have been exceeded by traditional fuels, charcoal and fuelwood, in both cities, and in both vented and unvented kitchens. For LPG and butane cookstoves, the results for $PM_{2.5}$ emissions were mixed. Nevertheless, the use of these two types of fuels still presented vast improvements in $PM_{2.5}$ emissions levels when compared to stoves utilizing fuelwood and charcoal. Any decrease in $PM_{2.5}$ would still be beneficial to health.

- **Comparisons with the US EPA Air Quality Index showed that traditional cookstoves using charcoal or fuelwood emit pollutants at levels that are very harmful to the health of every household member regardless of age group or gender.** While the one cooking will normally bear the brunt of health effects of the emissions, this does not spare other household members present or within close proximity during cooking from the harmful effects of HAP. Comparison of emission concentration measurements during the field emission tests with the US EPA AQI showed that charcoal and fuelwood emitted very unhealthy to hazardous levels of pollutants, especially $PM_{2.5}$ during kindling and cooking phases in both vented and unvented kitchens. Exposure would have tremendous health implications to all members of the household regardless of age or gender. This therefore makes it imperative that increasing access to cleaner cooking technologies and fuels should be given focused attention.

- **Electric stoves had the least negative impact on indoor air quality and health of not only those constantly tasked with preparing meals but also for all members of the household.** Based on WHO ERT standards and the US EPA AQI guidelines, electric stoves had the least negative impact on indoor air quality and the health of household members when compared with other fuels. Emissions from electric stoves were found to be very low to almost undetectable levels for the four pollutants: $PM_{2.5}$, CO, SO_2, and NO_2.

Cookstove and Cooking Fuel Efficiency

- **Electric stoves were found to have the highest thermal efficiency followed by LPG stoves. Some ICS were also observed (in a separate testing activity) to have higher thermal efficiency and could possibly be introduced as an alternative.** The thermal efficiency of the electric (coil) stove was 33.4% and the LPG stove 26.5%. The Mabaga Kalan (rocket stove), which was among the ICS with existing thermal efficiency tests included for comparison had a slightly higher thermal efficiency (27%) than the LPG stove tested in the laboratory. Other ICS also showed higher thermal efficiency than traditional cookstoves, which ranked the lowest among the stoves tested. Traditional stoves using charcoal had the lowest thermal efficiency at 5.2% followed by traditional stoves using fuelwood at 10.4%.

Estimates of Stove and Fuel Costs

- **Single and double burner gas (LPG) stoves have the highest up-front cost, for both Iloilo City and San Jose City.** Single burner LPG stoves utilizing an 11-kg LPG tank costs ₱4,071.75 ($93.37) in San Jose City and ₱4,295.75 ($80.75) in Iloilo City while a double burner LPG stove with the same LPG tank size will cost ₱4,967.00 ($93.37) in San Jose City and ₱5,445.86 ($102.37) in Iloilo City. These up-front costs of LPG stoves not only includes the burner, but also the cost of a filled LPG tank and its various accessories such as the valve regulator, hose, and hose clamp. Traditional cookstoves, usually made of cement, are the cheapest at ₱234.15 ($4.40) in San Jose City and ₱91.29 ($1.72) in Iloilo City.

- **Households exclusively using traditional cookstoves together with charcoal spend the most for fuel annually.** Because charcoal is the least efficient among the fuels tested in the laboratory, more charcoal is consumed when cooking the same amount of food as represented by boiling the same amount of water during the laboratory test as compared to the other fuel–technology combinations. In estimates using household survey data on cooking behavior, duration and price of the different fuels, and laboratory data on fuel consumption, it was estimated that households exclusively using traditional cookstoves with charcoal spend from ₱15,800.10 ($297.02) in San Jose City to ₱18,414.01 ($346.16) in Iloilo City for fuel. This is followed by LPG stoves, electric stoves, while traditional cookstoves using fuelwood remain the cheapest.

Outlook on Shifting

- **Switching from traditional cookstoves utilizing charcoal or fuelwood to modern cookstoves utilizing either LPG, butane or electricity will lead to significant reductions in PM$_{2.5}$ emission concentrations, especially in households exclusively using traditional cookstoves.** Field emission test results show large amount of reduction in PM$_{2.5}$ if a household switches from exclusively using traditional cookstoves that utilize charcoal and fuelwood to modern cookstoves using either LPG, butane or electricity. A switch from traditional charcoal-fueled cookstoves utilizing charcoal as cooking fuel, to modern cookstoves that utilize either LPG, butane, or electricity, can decrease PM$_{2.5}$ emission from as low as 60.04% to as much as 99.32%. A switch from traditional cookstoves utilizing fuelwood, to modern cookstoves, can decrease PM$_{2.5}$ emission by at least 84.38% to as much as 99.74%.

- **Switching from traditional cookstoves utilizing charcoal, to single (coil) electric stoves will lead to the highest annual cost reduction for the households.** Shifting from charcoal to a single (coil) electric stove can lead to annual household cost reduction from ₱12,588.78 ($236.65) in San Jose City to ₱14,535.75 ($273.25) in Iloilo City. This is because cooking with charcoal is more fuel-intensive and it has the lowest thermal efficiency among all the fuels tested in the laboratory. With the savings involved, households shifting from charcoal to modern cookstoves can expect to recover the up-front cost of cookstoves as early as less than a month if they shift to a single coil electric stove, or at most 5 months if they shift to a double burner stove with an 11 kg LPG tank.

- **Switching from traditional cookstoves to modern cookstoves will lead to GHG emission reductions.** If wood and charcoal are not sourced from plantations, a shift by households exclusively using fuelwood and charcoal to modern cookstoves would contribute to GHG emission reduction. A shift from fuelwood to LPG would result in GHG emissions reduction estimated at 14.6 tCO2e/yr or 17.7 tCO2e/yr if from fuelwood to electricity. Similarly, a shift from charcoal to LPG or electricity would result in some 90.5 tCO2e/yr or 102 tCO2e/yr of reduction, respectively.

- **Potential for use of ICS as an alternative to traditional cookstoves.** The promotion of the use of ICS can support the distribution and adoption of clean fuels and technologies. Especially for households practicing fuel stacking of traditional and modern cookstoves, ICS can act as a transitional cookstove for households preferring the use of charcoal and fuelwood for cooking due to cultural or other reasons. Energy efficient ICS can reduce fuel consumption as well as length of exposure to HAP. Though further studies are needed to ascertain performance standards of ICS available in the Philippines, the StovePlus Academy's initial study indicates that there are locally developed ICS technologies, with higher efficiency that can be promoted.

Barriers to Shifting to Modern and Clean Cooking Technology

- **The up-front cost of cookstoves, and the cost of fuel or additional cost to electricity are the primary barriers to shifting to modern cooking technologies identified by households.** From the survey results 54.3% of households in Iloilo City and 42.9% of households in San Jose City indicated cost as the primary barrier to shifting from traditional cookstoves to modern technologies using cleaner fuel. This is especially for those exclusively using traditional cookstoves. Furthermore, 43.6% of households in Iloilo City and 57.1% of households in San Jose City perceive that a shift to modern technologies will lead to increased expenses due to purchase of more expensive fuel (for LPG stove) or electricity (electric stove).

- **Households believe that exclusively using traditional cookstoves or stacking these in combination with modern cookstoves is advantageous.** The results of the study revealed that for 46.8% of households in Iloilo City and 20.0% of households in San Jose City, traditional cookstoves are generally more convenient to use. This is also connected with the belief that food cooks faster and leads to lower fuel expenses as indicated by 25.5% of households in Iloilo City and 17.1% of households in San Jose City.

- **Households believe that traditional cookstoves are safer to use than modern cookstoves.** Households have the belief that LPG stoves are prone to sudden or accidental explosions as expressed by 40.4% of households in Iloilo City. In San Jose City, while more households are already using LPG stoves, 8.6% of households also expressed this concern.

- **There is a need for an enabling policy environment on increasing access to clean cooking.** The national energy plan does not have specific policy direction toward increasing access to clean cooking, which is vital to set local initiatives into motion. National energy and environment agencies need to spearhead this initiative by setting the essential guiding principles. However, a HUC, such as Iloilo City, has also the authority, and may promulgate its own local directives and programs that can create awareness among households, incentivize local ICS developers, and encourage academes, financing institutions, etc. to engage in activities that would promote the use of and increase the market for clean cooking technologies and fuels.

6.2. Key Takeaways and Recommendations

The following key takeaways from the study can augment available information on the prevailing barriers and issues concerning the slow increase to access in clean cooking. The study can provide insight that can be translated into recommendations on how governments in general, and Iloilo City and San Jose City local governments in particular, as well as other energy policymakers, clean cooking technology investors, and other stakeholders can promote switching to cleaner cookstove alternatives and increase access to clean cooking. The following are key takeaways and recommendations:

1. **There is a need to create public awareness not only about the advantages of shifting to modern cooking technologies and fuels, i.e. thermal efficiencies and long-term cost savings, safety, and health implications but also of the advantages of proper ventilation to reducing indoor air pollution. Providing the public relevant information would help address the barriers to shifting to modern and cleaner cooking technologies.**

    •   There is a general lack of awareness among households and the general public about the negative impact of traditional cooking practices, particularly the use of fuelwood and charcoal as fuel for traditional cookstoves, on household air quality and the resulting impact of such practices on the health of household members, particularly, women, children, and the elderly. Awareness campaigns that highlight these will increase appreciation not only of the issues but also of the solutions and advantages of shifting to modern cookstoves. Relevant findings of this study could be captured into localized awareness campaign materials to facilitate understanding, retention and easy recall.

    •   Fuel preference remains one of the main drivers for maintaining traditional cookstoves. Fuel preference stem from not only for the unique flavor profile imparted by cooking with charcoal or fuelwood, cultivated through traditional and customary dishes, but also for the cost component. The convenience of being able to buy small and retail amounts of fuelwood and charcoal for the daily needs is more attractive to households and mislead them into thinking that they are saving more. By providing information concerning the various cooking technologies, their availability and comparative features, such as efficiency, safety and price, households are given the ability to make a better and more logical choice.

    •   Moving forward, it would be of great help to households if information about the advantages of proper household and kitchen ventilation is observed. Advisories to households and professionals, i.e., architects, interior designers, etc., can provide guidance on adjustments they can make on their current kitchen set-up and future kitchen designs to attain better indoor air quality.

2. **The practice of fuel stacking will persist, and improved cookstoves remain as an important option in many contexts. Technologies that bridge the gap between cultural needs and modern cooking alternatives should be part of the solution in increasing clean cooking access.**

    •   For the Philippines' context, ICS as an alternative to traditional cookstoves can address the continuous practice of fuel stacking. It can provide improved thermal efficiency, which will not only reduce the amount of fuel consumed but also the level of pollution caused by the burning of these fuels. ICS can likewise deal with cultural issues, such as flavor preference brought about by use of charcoal or fuelwood. However, because the development of ICS in the Philippines is still at its nascent stage, the performance of specific ICS models is still to be ascertained. Thus, it would be beneficial if performance and design standards for local ICS could be developed and given priority attention. Other modern technologies, such as solar cookstoves may also be considered, although supply sourcing, distribution, cost as well as market acceptance of the product may currently be barriers to the adoption of this technology. Nevertheless, it would be good to monitor the development of clean cooking technologies around the world for potential deployment into the Philippine local market in the future.

- Increasing recognition, and support to ICS technologies may also lead to better market choices. Government support to encourage commercial production of ICS may improve quality and reduce up-front cost and make these products more financially attractive to consumers. Other products that match cultural and preference concerns should likewise be promoted. For example, since a number of households already use rice cookers, the use of electric pressure cookers could possibly lessen the time and cost spent for boiling and tenderizing meat while keeping "taste or flavor" intact. The development of modern biomass fuels such as efficient biomass briquettes should likewise be considered as part of the technology development process. With financial support using appropriate financing models for both the manufacturing and end-use sectors, market expansion can be achievable. Microcredit, subsidized loans, and credit guarantees in the sector are some of the means to promoting access to financing.

3. **Increasing national and local government support is fundamental to the success of promoting a universal access to clean cooking by 2030.**

- Targeted and evidence-based, policies and plans to increase access to clean cooking technologies improves a country's performance in the drive for universal access to clean cooking. Policies on clean cooking should be integrated and be made an essential part of national energy plans. These plans should have measurable targets and allocated budgets to support awareness campaigns, financing for ICS development, manufacturing and distribution, as well as for implementing affordability gap measures that would allow low-income families to purchase modern cookstoves that are otherwise beyond their household budgets. To create a positive enabling environment, targeted phase out of subsidies for fuels used for cooking like kerosene, value-added tax and other taxes, and on policies on the imports or manufacturing of clean cooking stoves and cooking-related technologies, may be policy strategies to be contemplated. Moreover, it must be noted that the shift to clean cooking, particularly the use of electric stoves, would generate the most benefit when government also ensures that grid electricity is not fossil-fuel dependent. Thus, pushing for more renewable energy on the grid would be a complementary policy direction that government can pursue as well.

- The development of a comprehensive National Clean Cooking Country Action Plan that is built upon a credible baseline data of household needs, socioeconomic profiles, as well as regional and cultural distinctions would serve this purpose. A multisectoral approach, involving government and other stakeholders, such as, the health, environment, education, and energy departments, technology developers, supply chains, local government implementers and financial institutions can aid in this process to ensure the crafting of localized, and more comprehensive clean cooking solutions. Urban and rural areas would need different strategies and the National Clean Cooking Action Plan should cater to these differences. A study by the health department taking stock of the health and medical implications of indoor air pollution could also support this initiative.

4. **Each country must develop policies to address specific barriers uniquely their own.**

- This report shows that each country can have its own unique characteristics that must be addressed in a distinct way. This emphasizes the fact that there is no one-size-fits-all policy. While globally determined barriers, such as the UN-identified barriers are useful to note, not all are significantly true for all countries. Thus, policies must be developed to address issues or barriers specific to each country, considering local practices and preference, in order to provide a more effective approach to promoting and increasing universal access to clean cooking. Strategies should not only address technology but also promote local community support to encourage behavioral changes needed for a clean cooking transition.

- Local governments need not wait for a national policy framework before taking action. Access to clean cooking is a very local issue affecting the basic unit of any community, the family. Thus, LGUs should be encouraged to take the initiative and develop policies, guidelines and programs appropriate to their local needs, whether urban or rural. These may include policies, plans and projects to address and encourage use of clean cooking technologies, such as, cookstove standards and usage, guides to improving HAQ, awareness raising and capacity building, and investment incentives for expanding markets for clean cooking technologies. The role of women is critical in this sector. A participative approach involving women will not only enhance gender equality and promote inclusivity in the sector, but more likely, also influence the success of any program.

5. **The study in the Philippines can be refined and replicated in other localities within the country or among other DMCs.**

- This study has a number of limitations such as variables which the field enumerators were unable to take into account. Some findings need further study to understand the factors that produce seemingly unexpected results. The study, however, serves as a starting point that highlights the extent of how household cooking impacts indoor air quality and the need to take quick action, particularly because of its health implications. Considering the geographical and cultural diversity of communities in the Philippines, the study can be replicated in other regions across the country to look closely into unique local characteristics that may affect the introduction of clean cooking technologies. The study was also not able to go into the specifics of the health impacts. This is one area that can be included in further studies as a cross-cutting issue between environment and health agencies. Another area for further study would be specific to ICS, its current state and how it can be developed into a sustainable industry. Results could then be used as inputs to the preparation of local guidelines, rules and action programs suited to each locality.

- Similarly, other DMCs may learn from the findings of this study and can likewise develop their own local assessments to understand and prepare their respective national action plans. The Philippines experience emphasizes how all DMCs should look upon their own unique combination of barriers, which may hinder successful transition to universal access to clean cooking, and how important information is to raising awareness and understanding by LGUs and communities to local issues which then can encourage behavioral and policy changes.

6. **The key takeaways can serve as a guide for ADB in determining areas where it can significantly contribute to the global effort of achieving universal access to clean cooking by 2030**

From the output of the study, ADB can assess where it could contribute knowledge and resources in support of clean cooking access efforts, not only in the Philippines but also in other DMCs across the Asia and Pacific region. Specifically, this may be done through capacity building of government officials from DMCs, developing national action plans for clean cooking, pilot testing and demonstration of technologies, as well as in providing resources to both public and private cookstove and fuels stakeholders. It may also calibrate its approach to financing or initiate programs and partnerships directly focused on improving access to clean cooking.

Appendixes

Appendix 1: Description, Diagram, and Illustrations on the Filipino Kitchen Ventilation Categories

Ventilation Type	Kitchen Category	Description	Diagram	Example
VENTED	A	The cooking area is completely situated outdoors. Since it is exposed directly to open air and is mostly unobstructed, there is no need for exhaust pipes, fans and chimneys.	Living area / Open space	
VENTED	B	The cooking area is still in the house premises but facing a wall. The obstructing wall must not be shared with the living area. In effect, there is between one and two walls obstructing the free flow of air, but the rest of the cooking area is otherwise exposed.	Living area / Open space	
VENTED	B2	Same as Category B, but the obstructing wall is adjacent to the living area	Living area / Open space	

continued on next page

Appendix 1 continued

Ventilation Type	Kitchen Category	Description	Diagram	Example
VENTED	C	The cooking area is enclosed by three walls and is completely separated from the main living area. Screen meshes and wooden shutters count as "walls" also	Living area — Cooking area	
VENTED	C2	"Detached". Same as Category C, but at least one of the enclosing walls is shared with the living area.	Living area — Cooking area	
UNVENTED	D	The cooking area is located inside the main living area, and is enclosed by at least two obstructing walls (i.e. it is located in a corner of the living area)	Living area — Outside	

Legend: ——— Walls   - - - - Optional/additional walls   ■ Location of cookstove

Source: ADB. 2015. Promoting Sustainable Energy for All in Asia and the Pacific – Energy Access for Urban Poor. TA 8946. *Surveyors' Manual*. 2018. All Photos from Household Survey, ADB TA 8946, 2018.

Appendix 2: Survey Details

A surveyors' manual is drafted for the field survey questionnaire used by all participating surveyors for the proper implementation of protocol regarding field surveys. The surveyors' manual contains specific guidelines on how to properly record, encode, and process information from the respondents of the survey. All surveyors are briefed on the use of the questionnaire before the conduct of the field survey for both sites.

The survey and field emission tests were conducted on 201 households from 50 of 180 barangays in Iloilo City and 200 households from 25 of 38 barangays in San Jose City. Stratified random sampling was used in the selection of respondents. Of the barangays in San Jose City where surveys and field tests were conducted, seven were urban barangays and the remaining 18 were rural barangays. The barangays covered by the surveys and field tests in Iloilo City spanned the city's seven districts, with urban and rural barangays equally represented in the survey.

The questionnaire did not include much socioeconomic related data aside from the number and age of the household members. The survey aimed at providing information on the structural and occupancy aspects of households and included an inspection of the physical conditions and the state of kitchens inside the household, to identify structures that would indicate the level of ventilation that the respondents' kitchens had.

A section on respondents' perception regarding awareness on presence, availability, and concerns regarding switching to modern and cleaner cooking alternatives was included in the survey. Aside from these, the survey focused on gathering the following information:

- the type of cookstove and cooking fuel used;
- the duration of cooking, and the number of times in a day cooking was conducted;
- type of food prepared for each meal; and
- modes (retail, bulk) and frequency of purchase for fuels used by households.

References

ADB. 2013. *Energy for All: Increasing Access to Clean Cooking* (Brochure). https://www.adb.org/publications/ energy-all-increasing-access-clean-cooking.

Biolexis Multifuel Gasifier Stove Website: www.biolexis.com.ph

Cordes, 2011 as cited in World Health Organization. 2016. *Burning Opportunity: Clean Household Energy for Health, Sustainable Development, and Wellbeing of Women and Children.* Geneva. Pp. 130.

Energy Sector Management Assistance Program (ESMAP). 2020. *The State of Access to Modern Energy Cooking Services.* Washington, DC: World Bank. License: Creative Commons Attribution CC BY 3.0 IGO

GACC. 2014. Global Alliance for Clean Cookstoves. *The Water Boiling Test Version 4.2.3.: Cookstove Emissions and Efficiency in a Controlled Laboratory Setting.* 19 March. https://www.cleancookingalliance.org/binary-data/DOCUMENT/file/000/000/399-1.pdf.

Garg, A., K. Kazunari, and T. Pulles. 2006. *IPCC Guidelines for National Greenhouse Gas Inventories (Volume 2: Energy).* https://www.ipcc-nggip.iges.or.jp/public/2006gl/vol2.html.

GIZ. 2015. *ASEAN – German Technical Cooperation, Clean Air for Smaller Cities in the ASEAN Region.* Emission Inventory of Major Air Pollutants in Iloilo City (Final Report). Unpublished. https:// www.researchgate.net/publdication/326844171.

Government of the Philippines, Department of Energy. 2016. *Philippine Energy Profile 2016–2030.* https://www.doe.gov.ph/sites/default/files/pdf/pep/2016-2030_pep.pdf.

_____. 2019. *Philippine Energy Profile 2017–2040.* https://www.doe.gov.ph/pep/philippine-energy-plan-2017–2040.

Government of the Philippines, Department of Environment and Natural Resources. 2015. *Environmental Management Bureau: National Air Quality Status Report 2008–2015.* https://emb.gov.ph/wp-content/uploads/2015/09/1-Air-Quality-1.8-National-Air-Quality-Status-Report-2008-2015.pdf.

_____. 2018. *Philippine Forestry Statistics, 2018.* Forest Management Bureau. Manila.

Government of the Philippines – Department of Science and Technology. 2020. Climatological data provided by the Climatology and Agrometeorology Division of DOST-PAGASA. http://bagong.pagasa.dost.gov.ph/climate/climate-data. 4 April.

Government of the Philippines, Official Website of Iloilo City. www.iloilocity.gov.ph.

Government of the Philippines, Philippine Statistics Authority and Department of Energy. 2011. *Household Energy Consumption Survey 2011.* https://psa.gov.ph/hecs

_____. 2015. Census of Population and Housing: Highlights on Household Population, Number of Households, and Average Household Size of the Philippines https://psa.gov.ph/population-and-housing/node/69728

Government of the Philippines. 1991. Republic Act 7160: *The Local Government Code of the Philippines.* https://www.officialgazette.gov.ph/downloads/1991/10oct/19911010-RA-7160-CCA.pdf

Guinto, J. 2015. Anatomy of the PapaBrick Stove. September (unpublished). Accessed via http://www.drtlud.com/wp-content/uploads/2015/11/Anatomy-of-the-PapaBrick-Stove.pdf. 10 December 2019.

International Energy Agency. 2019. *Clean cooking access database.* https://www.iea.org/reports/sdg7-data-and-projections/access-to-clean-cooking. Accessed 28 April 2020.

_____. 2019. SDG 7: *Data and Projections (Access to affordable, reliable, and sustainable energy for all).* Flagship Report. November. https://www.iea.org/reports/sdg7-data-and-projections/access-to-clean-cooking. Accessed 11 September 2020.

_____. 2020. *Fossil vs biogenic CO2 emissions. IEA Bioenergy Technology Collaboration Programme* https://www.ieabioenergy.com/iea-publications/faq/woodybiomass/biogenic-co2/.

IEA, International Renewable Energy Agency, United Nations Statistics Division, World Bank, World Health Organization. 2019. *Tracking SDG 7: The Energy Progress Report 2019* (Data Annex). Washington, DC.

_____. 2020. *Tracking SDG 7: The Energy Progress Report.* World Bank: Washington DC.

Inzon, M.R.B.Q., M.V.O. Espaldon, et.al. 2016. Environmental Sustainability Analysis of Charcoal Production in Mulanay, Quezon, Philippines. *Journal of Environmental Science and Management.* 2016. pp. 93–100.

Ludovice, H.V. 2018. "Overview: Philippine Downstream Oil Industry" Presentation for the Department of Energy in the Energy Consumers and Stakeholders Conference with the theme: "E-Power Mo! Communicating Efficiency Across the Energy Sector." 24 April. Hotel Supreme, Baguio City, Philippines.

Manila Standard. 2018. Power rates in Iloilo highest among urban centers. https://manilastandard.net/news/national/280827/power-rates-in-iloilo-highest-among-urban-centers-.html. 18 November.

Oanh, Nguyen Thi. 2012. *Integrated air quality management: Asian Case Studies.* Edited by Kim Oanh N. T. CRC Press: New York.

Ortwien, Andreas and Militar, Jeriel G. 2015: *Use of Biomass as Renewable Energy Source in Panay. Final report.* Manila, Philippines: Deutsche Gesellschaft für Internationale Zusammenarbeit (GIZ) GmbH.

PowerPoint Presentation – Tracking SDG7: The Energy Progress Report. Lisbon, Portugal. 2 May 2018. A joint presentation of IEA, IRENA, UNSD, WB, and WHO. http://trackingSDG7.esmap.org.

PSMarketresearch. 2019. *Philippines Charcoal Market Research Report: By Application, Type - Industry Opportunity Analysis and Growth Forecast to 2030.* December.psmarketresearch.com.

Quinn, Ashlinn K., et al. "An Analysis of Efforts to Scale up Clean Household Energy for Cooking around the World." *Energy for Sustainable Development,* vol. 46, 2018, pp. 1–10., doi:10.1016/j.esd.2018.06.011.

Raish, J. n.d. Thermal Comfort: Designing for People. Edited by W. Land and A. McClain for The University of Texas at Austin, School of Architecture (Center for Sustainable Development) https://soa.utexas.edu/sites/default/disk/urban_ecosystems/urban_ecosystems/09_03_fa_ferguson_raish_ml.pdf.

Remedio, E.M., 2009. *An analysis of sustainable fuelwood and charcoal production systems in the Philippines: A Case Study. Criteria and Indicators for Sustainable Woodfuels: Case Studies from Brazil, Guyana, Nepal, Philippines and Tanzania.* Food and Agriculture Organization: Rome. http://www.fao.org/3/i1321e/i1321e00.pdf

SEforAll Energizing Finance Report Series: Taking the Pulse 2019, "Taking the Pulse of Energy Access in the Philippines" https://www.seforall.org/system/files/2019-12/Taking-Pulse-Philippines.pdf

Smith, K.R. n.d. Health impacts of household fuel use in developing countries. http://www.fao.org/3/a0789e09.htm.

SNV. n.d. Market Acceleration of Advanced Clean Cook Stoves in the Greater Mekong Sub-Region: End User Adoption Study (Vietnam). https://snv.org/cms/sites/default/files/explore/download/accs_end_user_adoption_study_snv.pdf.

StovePlus Academy. 2017. *Business Development for Improved Cookstoves and Innovative Fuels, 2017.* 4th Edition.

Sustainable Energy for All (SEforALL) and Catalyst Off-Grid Advisors. 2019. *Energy Finance: Taking the Pulse 2019* (Chapter Summary: Taking the Pulse of Energy Access in the Philippines). October. pp. 52-78. https://www.seforall.org/system/files/%20%202019-12/Taking-Pulse-Philippines.pdf.

United Nations. 2015. UN General Assembly: Transforming our world: the 2030 Agenda for Sustainable Development. 21 October. A/RES/70/1. https://www.refworld.org/docid/57b6e3e44.html.

_____. 2018. *Accelerating SDG 7 Achievement: Policy Briefs in Support of the First SDG 7 Review at the UN High-Level Political Forum* (Policy Brief #2: Achieving Universal Access to Clean and Modern Cooking Fuels, Technologies and Services. https://sustainabledevelopment.un.org/content/documents/17465PB2.pdf.

United Nations Development Programme. Sustainable Development Goals. Goal 7: Affordable and Clean Energy (Goal 7 Targets) https://www.undp.org/content/undp/en/home/sustainable-development-goals/goal-7-affordable-and-clean-energy/targets.html. Accessed on 28 May 2020.

_____. Sustainable Development Goals. Goal 7: Ensure access to affordable, reliable, sustainable and modern energy for all. https://unstats.un.org/sdgs/report/2016/Goal-07/.

_____. Sustainable Development Goals. https://www.undp.org/content/undp/en/home/sustainable-development-goals.html.

United Nations Economic and Social Commission for Asia and the Pacific. 2018. *Accelerating SDG 7 Achievement: Policy Briefs in Support of the First SDG 7 Review at the UN High-Level Political Forum* (Policy Brief #2: Achieving Universal Access to Clean and Modern Cooking Fuels, Technologies and Services. https://sustainabledevelopment.un.org/content/documents/17465PB2.pdf.

_____. 2020. *Accelerating SDG 7 Achievement in the time of COVID-19: Policy Briefs in Support of the High Level Political Forum of 2020* (Policy Brief: Advancing SDG 7 in Asia and the Pacific). https://www.unescap.org/sites/default/files/2020SDG7-POLICY-BRIEF-ASIA-PACIFIC.pdf) Accessed 10 June 2020.

United States Agency for International Development. 2017. *Analysis of the Charcoal Value Chain in Iloilo City* (Final Report). September. Prepared by the USAID Building Low Emission Alternatives to Develop Economic Resilience and Sustainability (B-LEADERS) Project.

US EPA. 2012. *Revised air quality standards for particles pollution and updates to the air quality index (AQI).* https://www.epa.gov/sites/production/files/2016-04/documents/2012_aqi_factsheet.pdf.

_____. 2014. *Air Quality Index - A guide to air quality and your health.* February. https://www3.epa.gov/airnow/aqi_brochure_02_14.pdf

_____. n.d. *AirNow Factsheet: Air Quality Forecasts and Observations.* https://www3.epa.gov/airnow/tvweather/airnow_factsheet.pdf.

World Bank. 2011. *Household Cookstoves, Environment, Health, and Climate Change : A New Look at an Old Problem.* Washington, DC. © World Bank. https://openknowledge.worldbank.org/handle/10986/27589 License: CC BY 3.0 IGO

_____. 2013. *Market Acceleration of Advanced Clean Cookstoves in the Greater Mekong Sub-region.* https://snv.org/project/market-acceleration-advanced-clean-cookstoves-greater-mekong-sub-region.

_____. 2014. CiDev.2019. *ERPA signed for Lao PDR Clean Cookstove Initiative.* 1 August. https://www.ci-dev.org/result-stories/erpa-signed-lao-pdr-clean-cookstove-initiative#:~:text=The%20Lao%20PDR%20Clean%20Cookstove%20Initiative%20aims%20to%20distribute%2050%2C000,the%20project%20on%20the%20ground.

_____. 2016. Death in the Air: Air Pollution Costs Money and Lives (Infographic). 8 September. https://www.worldbank.org/en/news/infographic/2016/09/08/death-in-the-air-air-pollution-costs-money-and-lives.

_____. 2017. *Fact Sheet: Efficient, Clean Cooking and Heating Program. An ESMAP/Sustainable Energy for All Partnership.* March. Washington, DC; ESMAP. 2017. *Nepal| Fostering Healthy Households through Improved Stoves.* https://www.esmap.org/node/57862#:~:text=The%20government%20of%20Nepal%20has,cookstoves%20to%20meet%20this%20goal. 29 March.

_____. 2018. Bangladesh: Healthier Homes through Improved Cookstoves. *Results Briefs.* https://www. worldbank.org/en/results/2018/11/01/bangladesh-healthier-homes-through-improved-cookstoves. 1 November.

_____. 2018. *Super-clean cookstove, innovative financing in Lao PDR project promise results for women and climate.* Feature Story. https://www.worldbank.org/en/news/feature/2018/04/20/clean-cookstove-innovative-financing-lao-pdr-project-promise-results-women-climate. 23 April.

_____. 2020. World Development Indicators (Population, total) https://data.worldbank.org/indicator/SP.POP. TOTL Last updated 1 July 2020.

_____. Pollution. https://www.worldbank.org/en/topic/pollution.

World Bank and Institute for Health Metrics and Evaluation. 2016. *The Cost of Air Pollution: Strengthening the Economic Case for Action.* Washington, DC: World Bank. License: Creative Commons Attribution CC BY 3.0 IGO.

World Green Building Council 2018. *Healthier Homes, Healthier Planet Guide.* London. https://www.worldgbc. org/sites/default/files/20181204_WGBC_Homes-Research-Note_FINAL_spreads.pdf.

World Health Organization. Ambient air pollution: Pollutants. https://www.who.int/health-topics/air-pollution#tab=tab_1.

_____. *WHO Indoor Air Quality Guidelines: Household fuel Combustion. (Recommendation 1: Emission Rate Targets (ERT))* https://www.who.int/airpollution/guidelines/household-fuel-combustion/ recommendation1/en/.

_____. *WHO Indoor Air Quality Guidelines: Household fuel Combustion (Review 2: Emissions of Health-Damaging Pollutants from Household Stoves).* Pp. 42.

_____. *Air Quality Guidelines: Household fuel combustion – Review 2: Emissions.*

_____. 2016. *Burning Opportunity: Clean Household Energy for Health, Sustainable Development, and Wellbeing of Women and Children.* Geneva, Switzerland. pp. 130.

_____. 2018. *Key Facts: Household Air Pollution and Health.* 8 May. https://www.who.int/news-room/fact-sheets/detail/household-air-pollution-and-health.

_____. 2020. Global Health Observatory: Proportion of Population with Primary Reliance on Clean Fuels and Technologies (%). https://apps.who.int/gho/athena/data/GHO/SDGPOLLUTINGFUELS?filter=COUN TRY:*;REGION:*&format=xml&profile=excel.